T0214530

Lecture Notes of the Institute for Computer Sciences, Social Informatics and Telecommunications Engineering 289

Adriana Compagnoni · William Casey ·
Yang Cai · Bud Mishra (Eds.)

Bio-inspired Information and Communication Technologies

11th EAI International Conference, BICT 2019
Pittsburgh, PA, USA, March 13–14, 2019
Proceedings

 Springer

Editors
Adriana Compagnoni
Stefens Institute of Technology USA
Hoboken, NJ, USA

William Casey
Carnegie Mellon University
Pittsburgh, PA, USA

Yang Cai
Cylab
Carnegie Mellon University
Pittsburgh, PA, USA

Bud Mishra
New York University
New York, NY, USA

ISSN 1867-8211 ISSN 1867-822X (electronic)
Lecture Notes of the Institute for Computer Sciences, Social Informatics
and Telecommunications Engineering
ISBN 978-3-030-24201-5 ISBN 978-3-030-24202-2 (eBook)
https://doi.org/10.1007/978-3-030-24202-2

This Springer imprint is published by the registered company Springer Nature Switzerland AG
The registered company address is: Gewerbestrasse 11, 6330 Cham, Switzerland

Preface

The 11th International Conference on Bio-inspired Information and Communications Technologies (BICT 2019) was held March 13–14, 2019, at Carnegie Mellon University in Pittsburgh Pennsylvania USA. BICT attracts researchers and practitioners in diverse disciplines that seek the understanding of key principles, processes, and mechanisms in biological systems and leverage those understandings in design, engineering, and technological applications. Past iterations of the conference have attracted significant contributions in direct bioinspiration (physical biological materials and systems used within technology) as well as indirect bioinspiration (biological principles, processes, and mechanisms used within the design and application of technology). This year, the scope expanded to include a third thrust: foundational bioinspiration (bioinspired aspects of game theory, evolution, information theory, and philosophy of science). Through foundational bioinspiration, we gain scientific and philosophical perspectives into the role of emergent bioinspired and biomimicry technologies and their wider implications and applications.

Biology offers an empirical and profound glimpse of dynamic stability, robustness, control, resilience, and survival. Accordingly, the application of biological research to systems and technology holds immense potential and reveals many technical challenges. We enjoyed a full two-day program featuring many fruitful discussions and presentations charting the current and future advances in bioinspired technologies.

The proceedings contain 14 accepted papers (acceptance rate of 52%). In addition to the accepted papers found here the 2019 BICT conference contained four special sessions: Human Machine Teaming (chaired by Ryan D. McKendrick, Northrup Grumman), Ethics in AI Applications in Industry (chaired by Thomson Nguyen, Kleiner-Perkins), and Re-Engineering Philosophy of Nature, Multiple Realisation and Natural Kinds (chaired by Paola Hernández-Chávez, University of Pittsburgh), and Nature and Games celebrating Bud Mishra's 60th birthday (chaired by Steven Massey, University of Puerto Rico). Additionally, an interactive music composition titled "Around the B-E-ES"! was presented by Jakub Polaczyk (NY Conservatory). The 2019 keynote speakers were: Sheri M. Markose (Professor of Economics at the University of Essex), Brian Skyrms (Distinguished Professor of Logic and Philosophy of Science, Economics, and Philosophy at University of California Irvine, and Professor of Philosophy at Stanford University), Michael Lotze (Department of Surgery, Immunology, and Bioengineering), and Bill Novak (Software Engineering Institute, Carnegie Mellon University).

June 2019

Adriana Compagnoni
William Casey
Yang Cai
Bud Mishra

Conference Organization

Steering Committee

Imrich Chlamtac	University of Trento, Italy
Jun Suzuki	University of Massachusetts, Boston, USA
Tadashi Nakano	Osaka University, Japan

Organizing Committee

General Chair

Bud Mishra	New York University, USA

Technical Program Committee Co-chairs

William Casey	Software Engineering Institute, Carnegie Mellon University, USA
Yang Cai	Carnegie Mellon University, USA
Jun Suzuki	University of Massachusetts, Boston, USA
Eric Hatleback	Carnegie Mellon University Software Engineering Institute, USA

Publicity and Social Media Chair

Bilal Khan	University of Nebraska-Lincoln, USA

Workshops Chair

Aftab Ahmad	The City University of New York, USA

Sponsorship and Exhibits Chair

Mohammad Upal Mahfuz	University of Wisconsin-Green Bay, Green Bay, USA

Publications Chair

Adriana Compagnoni	Stevens Institute of Technology, USA

Local Chair

Linda Canon	Software Engineering Institute, Carnegie Mellon University, USA

Web Chair

Tadashi Nakano	Osaka University, Japan

Conference Manager

Andrea Piekova   EAI

Technical Program Committee

Shih-Hsin	Chen Cheng-Shiu University, Taiwan
Adam Noel	University of British Columbia, Canada
Neil Walkinshaw	University of Leicester, UK
Jun Hakura	Iwate Prefectural University, Japan
Petra Hofstedt	Brandenburg University of Technology Cottbus-Senftenberg, Germany
Hironori Washizaki	Waseda University, Japan
Li-Wei Chen	National Kaohsiung Normal University, Taiwan
Shaukat Ali	Simula Research Laboratory, Norway
Andrew Schumann	University of Information Technology and Management in Rzeszow, Poland
Georgios Sirakoulis	Democritus University of Thrace, Greece
Md Abdur Rahman	Federation University, Australia and AIUB
He Peng	University of Electronic Science and Technology of China
Hiroaki Fukuda	Shibaura Institute of Technology, Japan
Paul Leger	Universidad Católica del Norte, Chile
Tomohiro Shirakawa	National Defense Academy of Japan
Dariusz Mrozek	Politechnika Śląska, Poland
Liguo Yu	Indiana University South Bend, USA
Preetam Ghosh	Virginia Commonwealth University
Vijender Chaitankar	NIH
Cem Sahin	MIT Lincoln Laboratory
Kei Ohnishi	Kyushu Institute of Technology
Michael Mayo	U.S. Army ERDC
Gang Qu	UMD
Krishna Venkatasubramanian	WPI
Chih-Wei Huang	National Central University, Taiwan
Hyun-Ho Choi	Hankyong National University
Muhammad Rizwan Asghar	The University of Auckland
Kyung Sup	Kwak Inha University
Raphael Machado	Inmetro
Saori Iwanaga	Japan Coast Guard Academy
Yifan Chen	South University of Science and Technology of China
Chun Tung Chou	The University of New South Wales
Soichiro Tsuda	University of Glasgow
Emanuela Merelli	University of Camerino
Sjouke Mauw	University of Luxembourg

Chih-Yu Wang	Research Center for Information Technology Innovation, Academia Sinica
Jose Morales	Software Engineering Institute, Carnegie Mellon University
Chonho Lee	Nanyang Technological University
Maurizio Porfiri	New York University Polytechnic School of Engineering
Munehiro Takimoto	Tokyo University of Science
Yusuke Nojima	Osaka Prefecture University
Parisa Memarmoshrefi	University of Goettingen
Vincent Cicirello	Stockton University
Yi Ren	UMASS Boston
Stanislav Tsitkov	Columbia University
Pruet Boonma	Chiang Mai University
JungRyun Lee	Chung-Ang University
Yasushi Kambayashi	Nippon Institute of Technology
Hirotake Yamazoe	Ritsumeikan University
Krzysztof Pancerz	University of Rzeszow
Yukio Gunji	Waseda University
Eric Hatleback	Carnegie Mellon University Software Engineering Institute
Thomas Schmickl	University of Graz, Austria Alan Davy Waterford Institute of Technology
Victor Erokhin	CNR-INFM and Department of Physics, University of Parma
Elena Zaitseva	University of Žilina
Behzad Moshiri	University of Tehran, Iran and University of Waterloo, Canada
Taichi Haruna	Kobe University
Qiang Liu	University of Electronic Science and Technology of China
Yasir Malik	University of Alberta
Kazuto Sasai	Tohoku University
Sergio Segura	University of Seville
Pedro Rangel	Henriques Universidade do Minho, Dep. de Informatica
Reiji Suzuki	Nagoya University
Kevin Pilkiewicz	Scientist, U.S. Army Engineer Research and Development Center
Raphael Machado	Inmetro

Contents

Cheating the Beta Cells to Delay the Beginning of Type-2 Diabetes Through Artificial Segregation of Insulin

Huber Nieto-Chaupis[(✉)]

Universidad de Ciencias y Humanidades UCH,
Av. Universitaria 5175, Lima39, Los Olivos, Peru
hnieto@uch.edu.pe

Abstract. In this paper, we focus in an artificial mechanism to detain the beginning of the type-2 diabetes disease in those identified patients which might to be developing a phase of prediabetes. From purely electrical interactions or Coulomb forces between a deployed nano sensor around of beta cells and Calcium^{2+} ions, we propose an artificial entrance of Calcium ions inside the beta-cells allowing them to segregate insulin. The electrical interactions between positively charged insulin inside beta cells is the main assumption of this paper. The permanent segregation of insulin fits well inside of the architecture of advanced networks engineering that contemplates the usage of a bio cyber interface. Therefore, the artificial releasing of granules with repulsive electric forces of insulin becomes a manner to cheat beta cells. This might be also seen as an option to avoid the intake of prediabetes y diabetes pharmacology for large periods. Although the view of this work is theoretical and prospective, it is based entirely in closed-form physics equations that sustain the main claim of this paper: electric interactions driven by charged nano particles would be a window to stop the progress of diseases based on the induced or spontaneous deficit of proteins, hormones and cells that are crucially needed to maintain the human homeostasis.

Keywords: Diabetes · Beta cells · Insulin

1 Introduction

Nowadays, type-2 diabetes becomes a worldwide issue that requires efforts from all angles of the scientific procedures [1]. Essentially, the disease appears due to the anomalous functionality of the beta cell that is not capable to segregate insulin. It has well-defined consequences against the homeostasis in human beings. In fact, the onset of the disease might not have signals, however there is a progressive and in some cases might be called aggressive behavior of the disease in the sense of the degradation of key organs such as kidneys, heart and the vascular system for large terms. Commonly, one sees the existence of two well-defined phases: the prediabetic and diabetic. Normally when the patient presents

© ICST Institute for Computer Sciences, Social Informatics and Telecommunications Engineering 2019
Published by Springer Nature Switzerland AG 2019. All Rights Reserved
A. Compagnoni et al. (Eds.): BICT 2019, LNICST 289, pp. 1–13, 2019.
https://doi.org/10.1007/978-3-030-24202-2_1

lectures of fasting glucose test above 130 mg/dL, the endocrinologist gives a diagnostic of type-2 diabetes. Thus, the diabetic patient has crossed the line that separates the prediabetic and diabetic states and is strongly recommended to intake among various types of pharmacology the well-known metformin. Therefore, patients might to continue with this intake for large periods, however it does not guarantees to locate to the patient away from risk situations [2].

The deficit to segregate insulin is a fact entirely related to the well-known beta cells that are responsible to segregate in an unstoppable manner the granules of insulin. The action to segregate insulin has as previous action the mechanism of depolarization by which the beta cell creates a voltage that allow the entrance of Calcium^{2+} ions [3]. Once these ions reach to enter inside the cell, the insulin is segregated. This process is fully spontaneous and only requires charged ions as external agents to achieve the segregation of insulin. The why beta cells cannot segregate insulin then might be derived from

- the Calcium ions are far away from the beta cells, so there is not physical contact between them, thus insulin remains inside the beta cells and all of them acquire a sterile state.
- the Calcium ions achieve to enter inside the beta cells but are not enough to push out the insulin, so the problem is seen as a lack of volume of Calcium ions [4].

Based on these facts, emerges the idea of cheating in the sense of reconfigure the scenario of insulin segregation but using artificial methods. One of them might be entirely in the territory of physics that can explain the processes of insulin production. Certainly the application of physics principles and subsequently equations, requires the understanding of the phenomenon. Once it is done we can use all those equations that fits well with the dynamics and interactions among the elements of system.

Because in one hand we have the fact of the entrance of Calcium ions through the beta cells, then the physical property that emerges as a key piece to achieve the segregation of insulin becomes the sign of the electric charge of such ions. Thus, we identified that occurs to some extent the processes of electric interactions either outside or inside the beta cell. Although literature have not explained in all the involved processes that are done because the electric charge [5], in this work we claim that the entrance of Calcium^{2+} inside the beta cell involves a scenario of Coulomb interactions essentially between the insulin granules and Calcium ions.

In this manner, while from the physics view we see that the segregation is a purely dynamical situation, then it should be due to the physical causes such as the electric forces between Calcium ions and the electric charge of the insulin still inside of the beta cell. In this paper we propose a mechanism that would inhibit to those prediabetic patients to be apart from a possible acquisition of the disease as well as to avoid the intake of pharmacology. To do that, we present a physics-based scheme that allows us to cheat beta cells targeting their segregation of granules of insulin through the presence of nano device that would controls the volume of segregated granules.

Second section is entirely devoted to the derivation of the main equations that are responsible to the production of insulin. Third section presents the results of paper, while in fourth section the proposal to engage the mechanism of insulin segregation is given. Conclusion of paper is drawn in last section.

2   The Physics Foundations for Cheating

Inside the context of insulin production, we define "cheating" as the action that allows us to stimulate with artificial methods the production of insulin with external nano devices that contain a electric charge. In this view, these proposal would demand the usage of physics equations.

In order to derive equations that would describe the artificial segregation of insulin by the beta cells we assume that

- beta-cell contains a volume,
- beta-cell has a certain density of granules of insulin inside,
- insulin has an electric charge whose sign might not be well defined (For example, Pizarro-Delgado [6] has shown the releasing of insulin in Langerhans islets once were depolarized through an experimental model.),
- insulin would react by electric interactions around,
- beta-cell carry out the processes of depolarization,
- the entrance of Calcium in the beta-cell is a cause for the segregation of insulin.

According to experiments done in the past, insulin is released from beta cell as a response to the processes of depolarization, fact that would suggests that the releasing is strongly based on the electrodynamics of ions more than purely biochemical processes.

Consider ρ_{Ca} as the density of Calcium 2+ ions moving inside the beta cell. Then there is a linear relationship between these ions and the segregation of insulin:

$$\rho_{Ca} = \gamma \rho_I \qquad (1)$$

where ρ_I the density of granules of insulin leaving the beta cell and γ a constant of proportionality. When both densities are electrically charges each other then both have a total charge in the sense that

$$Q_{Ca} = \int \rho_{Ca} dV \qquad (2)$$

$$Q_I = \int \rho_I dV, \qquad (3)$$

thus we can propose a repulsion or attraction force between them by assuming Coulomb interactions [7]:

$$\mathcal{F} = \frac{1}{4\pi\epsilon_0} \frac{Q_{Ca} Q_I}{|\mathbf{x} - \mathbf{x'}|^2}, \qquad (4)$$

and the energy that the system expends to apart from each other that is translated as the required energy to move out granules of insulin is given by:

$$W = \int \mathcal{F} d\mathbf{x} = \frac{1}{4\pi\epsilon_0} \int \frac{\mathcal{Q}_{Ca} \mathcal{Q}_{I} d\mathbf{x}}{|\mathbf{x} - \mathbf{x}'|^2}. \tag{5}$$

Clearly this energy is not controllable but while we can control the dynamics of the Calcium ions, then the system is under control. Therefore to impose controllability on the Calcium ions, we need an external agent that acquires the role of exert movement to the Calcium ions located around the beta cells. Prospective and ideas about the deployment of nano devices is seen in [8].

The next step is to represent in a closed-form a solution for each density. In the case when these solution are time-dependent then a good choice is the usage of the diffusion equations as follows

$$\frac{\partial \rho_{Ca}(s,t)}{\partial t} = D_1 \nabla^2 \rho_{Ca}(s,t) \tag{6}$$

$$\frac{\partial \rho_I(s,t)}{\partial t} = D_2 \nabla^2 \rho_I(s,t) \tag{7}$$

where D_1 and D_2 the diffusion constants for both cases. To note that the usage of the diffusion equation demands to assume that the densities have explicit dependence on the time.

2.1   Naive Derivation of the Releasing of Granules of Insulin

We define ρ_R as the density of the released granules of insulin. In virtue of Eq. 1 we write down a generalized relation involving densities

$$\rho_R = \gamma \mathcal{F} \Delta V \Delta T \tag{8}$$

that express the fact that the amount of released granules of insulin depends directly on the constant γ as well as the electric force between the charges inside the beta cell and the charges outside the cell. The exerted granules would also depend on the volume of the cell and the time that takes the electric interaction. Equation 8 becomes the fundamental equation of balance of segregation of insulin by the beta cell. Commonly, classical pharmacology as metformin induces the segregation and it can take a time to reach an optimal value. Thus it the optimal releasing of insulin would depend entirely on time after the biochemical reactions have ended. Thus we can understand this as

$$\frac{d\rho_R}{dt} = \Delta V \Delta T \frac{d(\gamma \mathcal{F})}{dt} + (\gamma \mathcal{F}) \frac{d(\Delta V \Delta T)}{dt} = 0 \tag{9}$$

to find their maximum and minimum values. A straightforward algebra with $d(\Delta V \Delta T) - \Delta V dt = 0$ yields that

$$\gamma \mathcal{F} = \mathrm{Exp}\left[-\frac{t}{\Delta T}\right] \tag{10}$$

When the result is inserted in Eq. 8 we get

$$\rho_R = \text{Exp}\left[-\frac{t}{\Delta T}\right]\Delta V \Delta T \tag{11}$$

turning now to a more realistic interpretation of Eq. 11 we adjudicated to ρ_R the following meaning: the number of granules of insulin per unit of volume and per unit of time, thus

$$\rho_R = \frac{N_R}{\Delta V \Delta T} = \text{Exp}\left[-\frac{t}{\Delta T}\right]\Delta V \Delta T \tag{12}$$

thus the N_R can be written as

$$N_R = \text{Exp}\left[-\frac{t}{\Delta T}\right](\Delta V)^2 (\Delta T)^2 \tag{13}$$

and with the frequency $\omega = \frac{1}{\Delta T}$ and $N_0 = \Delta V^2$ we have

$$N = N_0 \frac{\text{Exp}\left[-\omega t\right]}{\omega^2}. \tag{14}$$

The frequency is clearly understood as the inverse of the periods by which the releasing is done. These periods are determined by the electric interactions between the granules of insulin and the Calcium 2+ ions.

2.2 Full Usage of Physics: Electrodynamics and Diffusion Equations

Inspired on Eq. 8 that encloses the physics of the action of releasing the insulin granules, we consider the electric force as a possible cause to produce displacements and dynamics associated to the charge density. In this manner Eq. 1 can be rewritten as

$$\rho_R = \gamma \rho_{Ca}(\mathbf{r}, t)\rho_I(\mathbf{r}, t)d^3\mathbf{r}dt \tag{15}$$

where the integration runs over the volumes containing the total charges as $q = \int \rho dV$. We remark the presence of γ that can be perceived as the quantity that is human controllable and to some extent depends on the external physical variables.

In this manner γ is to be defined entirely inside of the territory of the electromagnetic propagation as we shall see along the next sections. In fact, external physical variables might have capabilities to regulate the electric force between the Calcium ions and their effect on the granules of insulin inside the cell. The next steps in this analysis require the following assumptions:

- the variation of the Calcium ions density would have consequences on the cells;
- the density of Calcium ions is controllable by an external device;
- the external device is dependent on the electromagnetic pulses monitored by a bio cyber interface;

- the presence of nano devices around the beta cells do not alter the biochemical reaction inside the Langerhans islets.

In virtue to the previous assumptions we can propose up to 2 well established scenarios:

Scenario I

This first scenario considers that γ is a constant and the density of released granules is in essence dependent on the Coulomb-like character of the interactions that take place inside the Langerhans islets,

$$\rho_R(\omega) = \frac{dN_R}{dV} = \gamma \rho_{Ca}(\mathbf{r}, t)\rho_I(\mathbf{R}, t) \tag{16}$$

so that the number of released granules is derived in a straightforward manner and it reads as

$$N_R = \gamma \int \rho_{Ca}(\mathbf{r}, t)\rho_I(\mathbf{R}, t)dV. \tag{17}$$

Therefore we can estimate the number of granules from the solutions of the diffusion equation for both charge densities. It should be noted that whereas $\mathbf{r}$ describes the vector that defines the position of the Calcium 2+ ion, $\mathbf{R}$ denotes the position of the insulin granule, so the integration runs over the volume $d^3\mathbf{R}$ that would enclose the granules. Concretely, these granules are all those that are being displaced by the Coulomb force exerted by the Calcium ions. Clearly, the repulsion is the case that fits with the action of segregation. Thus, as long as both charge distributions have same electric charge then the repulsion is imminently expected. Subsequently, the exact expression that might be able to estimate the number of released insulin granules is written as

$$N_R = \gamma \int \left[\int \rho_{Ca}(\mathbf{r}, t)d^3\mathbf{r} \int \rho_I(\mathbf{R}, t)d^3\mathbf{R} \right] dV \tag{18}$$

Scenario II

Because the dependence on a frequency as incorporated in Eq. 14 is justified on the basis that the Calcium ions dynamics that would depend of the strength of the electric force intensity between it and an external nano device, then in virtue of Fig. 1 the full electrodynamics of system might be described in terms of the electromagnetic pulses [9] that are sent by the external bio cyber interface to the nano device. Clearly, is the nano device is also a charged object then it exerts electric force either repulsion or attraction on the Calcium ions.

Thus γ encloses information about the electromagnetic pulse intensity. In this manner $\gamma \to I(\omega, t)$ denoting the well-known function of classical radiation that is sent by the bio cyber interface. Therefore we relate out that the induced releasing of granules of insulin depends entirely on the displacements of Calcium ions. On the other side the charge density of these ions are then fully dependent on the frequency of the pulses ω. Therefore the number of released granules is written in a straightforward manner as:

$$\rho_R(\omega) = \frac{N_R}{dV} = I(\mathbf{r}, \omega, t)\rho_{Ca}(\mathbf{r}, t)\rho_I(\mathbf{R}, t) \tag{19}$$

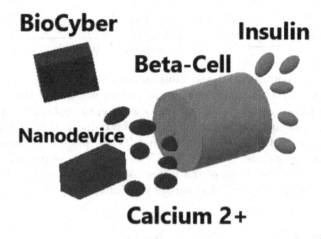

Fig. 1. Schematic representation of the action of cheating beta cells with Coulomb electric forces. The bio cyber interface sends an electromagnetic field characterized by a frequency ω to nano devices whose role is that of exert electric force to the Calcium 2+ ions to guarantee their entrance the beta cell and therefore to be able to sengregate granules of insulin.

Thus, the net number of released insulin granules by one beta cell is governed by the following master equation:

$$N_R(\omega) = \int I(\mathbf{r}, \omega, t)\rho_{Ca}(\mathbf{r}, t)\rho_I(\mathbf{r}, t)dV\,dt. \tag{20}$$

Generalizing previous equation for N cells, we have the total for these cells

$$N_R(\omega) = \sum_q \int I(\mathbf{r}, \omega, t)\rho_{Ca}(\mathbf{r}, t)\rho_{I,q}(\mathbf{r}, t)dV_q dt. \tag{21}$$

3 Results and Simulations

To solve the charge densities through the usage of the diffusion equation $\frac{\partial}{\partial t}\rho(\mathbf{r}, t)$ $= D\nabla^2\rho(\mathbf{r}, t)$ where D the diffusion's constant. For the particular case we search for a methodology that offers a closed-form solution for both the Calcium ions and the insulin granules inside of the beta cells by using standard closed-form techniques for solving the diffusion equations we have assumed that nanodevice and beta cells have a cylindric geometry. Therefore with all this we arrive to analytic solutions as shown below,

$$\rho_{Ca}(r, z, \theta, t) = \sum_{\ell=-\infty}^{\infty} \frac{e^{-\lambda D_C(t-t_0)}\mathrm{Sin}\left(\frac{n_1\pi z}{z_1}\right)\mathrm{Sin}\left(\frac{n_2\pi\theta}{\theta_1}\right)J_\ell(qr)}{J_\ell(qr_1)J_\ell(qr_2)} \tag{22}$$

$$\rho_{\rm I}(r,z,\theta,t) = \sum_{m=-\infty}^{\infty} \frac{e^{-\lambda D_I(t-t_0)} \cos\left(\frac{n_1\pi z}{z_2}\right) \cos\left(\frac{n_2\pi\theta}{\theta_2}\right) J_m(qr)}{J_m(qr_3) J_m(qr_4)}, \qquad (23)$$

where D_C and D_I correspond to the diffusion constants for the Calcium ions and insulin respectively. The quantities r_1, r_2 and r_3, r_4 denote the radial distances in both cylindrical geometries of ion and the receptor beta-cell. All of them were used as boundary conditions for solving the radial part of the differential equations. Below in Fig. 2 both Eqs. 22 and 23 are plotted for $lambda D(t - t_0) \gg 1$ so the exponential is nearly to 1. The apparition of peaks is perceived as a consequence of the usage of the integer-order Bessel function being these functions the solution to the radial part of the diffusion equation.

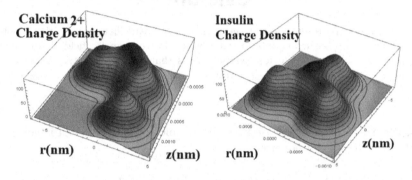

Fig. 2. The 3D plots of Eqs. (22) and (23). Left side: the case where Calcium 2+ by assuming roughly a cylindric flux of Calcium. Right side: the insulin as function of radius and length of the cylindric shape of beta cell.

3.1 The Description of the Electromagnetic Pulse

The to model $I(\mathbf{r},\omega,t)$ accurately, it is required the usage of the Friis equation. We turn now to explicitly define the electromagnetic pulse which is assumed to be emitted by the bio-cyber interface and received by the nanodevice. Intuitively one can assume that the received signal might be modeled at first instance by the Friis-like equation [10] $I(\omega) = I_i(\omega)\frac{A_n A_j}{(d\lambda)^2}$ where A denotes the effective distances whereas d the average separation distance and λ corresponding to the emitted pulse. Therefore we focus on the form of $I_i(\omega)$ as function of the emitted intensity by the bio-cyber interface and can be write down as

$$I(\omega,r,\theta,t) = \frac{A_n A_i}{(d\lambda)^2} e^{-i(\omega t - \frac{2\pi}{\lambda} r)} \eta \frac{|I_0|^2}{8\pi^2 r^2} \sin^3\theta \qquad (24)$$

that it does not depends on the z. To note $|I_0|$ denotes the pulse intensity. Therefore the full flux of released insulin granules per volume and time units dictated by Eq. 19 and together with the solutions Eqs. 22 and 23 becomes entirely dependence from the frequency ω and it is given for $t_0 = 0$ as:

$$\rho_{\mathrm{R}}(\omega) = \int (AI_0)^2 e^{-i(\omega t - \frac{2\pi}{\lambda}r)} \eta \sin^3\theta \times e^{-\lambda(D_C + D_{\mathrm{IN}})t}$$

$$\times \frac{\mathrm{Sin}\left(\frac{n_1\pi z}{z_1}\right) \mathrm{Sin}\left(\frac{n_2\pi\theta}{\theta_1}\right) \mathrm{Cos}\left(\frac{n_1\pi z}{z_2}\right) \mathrm{Cos}\left(\frac{n_2\pi\theta}{\theta_2}\right) J_m(qr)J_\ell(qr)}{(d\lambda)^2 8\pi^2 r^2 J_\ell(qr_1)J_\ell(qr_2)J_m(qr_3)J_m(qr_4)} dV dt. \quad (25)$$

resulting that exists there a direct dependence on the pulse intensity and the areas due to the Friis-like approximation.

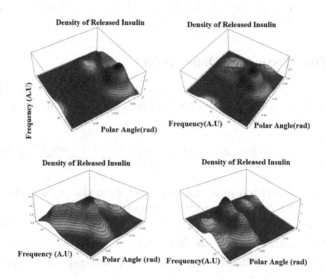

Fig. 3. Plots of Eq. 25. The charge density of the released granules of insulin as function of the frequency (arbitrary units) and the polar angle. Top left panel: the released might be seen as discrete packets, one prominent packet is observed. Top right panel: Secondary packets are segregated apart from the main one. Bottom left panel: packets are segregated as consequence of the increasing of the frequency of pulses. Bottom right panel: frequency can tune the segregation as seen in figure.

The modeling of the pulse intensity is inspired on the one derived from the half-wave dipole approximation [11]. However one can anticipate a clear limitation of the present model as to use intensities that actually would hazard human tissues. So the usage of those intensities over the range of small radiation exposure deserves a wide treatment. Therefore the intensity I_0 is assumed to be fixed and with the lowest values by allowing an acceptable signal reception by the nanodevice [12–14]. We can see that the integration over t, z, and θ turns out to be in a straightforward manner. The integration in the radial component requires a different analysis. Thus as a first task we can anticipate the dependence of the insulin releasing by the beta-cells with respect to the frequency of the arriving pulses to the nanodevice.

In Fig. 3 top and bottom: left and right panels are displayed the 3D distributions from Eq. 25 where the density $\rho(\omega)_R$ is plotted against the polar angle and frequency. Here are displayed the number of released granules of insulin as function of θ and frequency. Clearly the releasing of insulin is given through packets fact that is perceived in the peaks as consequence of the presence of the sinusoid functions sin and cos and the Bessel functions. One can also see that this density of granules of insulin decreases with the frequency of communication between the cyber-human interface and nano device [15,16]. However, one can see the presence of peaks for certain values of frequencies that is interpreted as possible events of hyperinsulinism.

3.2 The Time-Frequency Dependence

The temporal contribution to the integration Eq. 25 can be done quickly. For the sake of simplicity we shall consider the real part so that the integration yields a closed-form expression as:

$$\rho_R(\omega) = e^{-(D_C+D_{IN})T}\frac{[(D_C + D_{IN})\cos\omega T - \omega\sin(\omega T)]}{(D_C + D_{IN})^2 + \omega^2} \qquad (26)$$

Equation 26 can be written in a most compact form that would yields another morphologies of the curves of insulin segregation by the beta-cells. Thus, we define $\Delta = \mu\sin\gamma$ and $\omega = \mu\cos\gamma$ that implies that $\mu = \sqrt{\Delta^2 + \omega^2}$ where $\Delta = (D_C + D_{IN})$ so in this manner we have that $\gamma = \tan^{-1}\left[\frac{(D_C+D_{IN})}{\omega}\right]$ and the frequency-dependence of the insulin granules per beta-cell can be written as

$$\rho_R = e^{-\Delta t}\left\{\frac{(\sqrt{\Delta^2 + \omega^2})\sin\left[\text{Tan}^{-1}\left(\frac{\Delta}{\omega}\right) - \omega T\right]}{\Delta^2 + \omega^2}\right\} \qquad (27)$$

In Fig. 4 the density of released insulin as function of time and frequency by following Eq. 27 are plotted. We present two cases where the releasing is done through the emission of well-defined packets (left side) whereas the scenario where the emission is limited to some values of frequencies [17] is seen in right side. The fact that the existence of a oscillatory behavior triggers the idea that the electric energy is governed by a full process of oscillatory dynamics by which a Hamiltonian or Lagrangian can be adjudicated. Therefore the process of cheating might be governed by aspects of energy between the Calcium ions and the insulin inside the beta cells.

3.3 When the Cheating Fails

Of course, not all electric interactions can be successful since the population of Calcium 2+ ions and granules of insulin belong to any well-established statistical distributions. Although this is beyond of the scope of this work, the

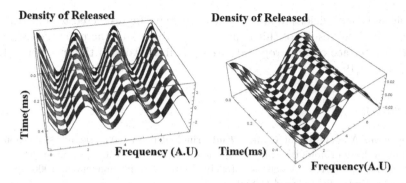

Fig. 4. The insulin density as given in Eq. 27 when $(D_C + D_{IN}) \approx 1.5$. The resulting surfaces contain a successive sequence of peaks for different frequencies and periods as consequence of the usage of the integer-order Bessel functions.

artificial process for cheating might fail for certain times. Thus ρ_R is null when $\sin\left[\mathrm{Tan}^{-1}\left(\frac{\Delta}{\omega}\right) - \omega T\right] = 0$. As consequence of this, we have that $\mathrm{Tan}^{-1}\left(\frac{\Delta}{\omega}\right) - \omega T = 0$. Then we can arrive to $\Delta = \omega\mathrm{Tan}(\omega T)$, as seen before, $T = t - t_0$ is the difference between the first segregation and the subsequent ones, thus one expects that for large times there is also a substantial probability to expect that the expected cheating might fail. Although the Coulomb forces are imminently of character deterministic, the processes to push out granules of insulin from the beta cell might be a process entirely governed by stochastic events. In fact, we can propose a probability distribution function that regulates this segregation and it reads as: $\mathcal{P}(r,\omega) = P_0\mathrm{Exp}\left[-\frac{r^2\omega\mathrm{Tan}(\omega T)}{\Delta}\right]$, where P_0 a normalization constant and r the radial distance between the granule of insulin and the Calcium 2+ ion.

4   Conclusion

Along this paper we have presented a methodology based on physics laws that would allows us to cheat beta cells in order that all of them can acquire capabilities to segregate insulin. This proposal relies entirely on the presence of a nano device that exert electric force on the beta cells in order that the segregation of granules of insulin fulfill the requirement of stabilize the patient to do not surpass the line that separates of being or not being a diabetic patient. Nano devices or nano robots [18] are expected to play a relevant role to improve the quality-of-life of diabetic patients. In the present paper we have obtained simulations that demonstrate the prospectiveness of the proposed method that would cheat beta cells through electrical interactions. Although the procedure of cheating is not efficient in all, the cheating might be restricted to some values of frequency. The results of this paper would indicates us the potential usage of bio cyber interface as the proposed inside the framework of the Internet of Bio Nano Things [19].

References

1. American Association of Clinical Endocrinologist: The American Association of Clinical Endocrinologists Medical Guidelines for the Management of Diabetes Mellitus: The AACE system of intensive diabetes self-management-2000 update. Endocr. Pract. **6**(1), 43–84 (2000)
2. Kaarisalo, M.M., et al.: Diabetes worsens the outcome of acute ischemic stroke. Diab. Res. Clin. Pract. **69**(3), 293–298 (2005)
3. Yoshida, M., et al.: Regulation of voltage-gated K+ channels by glucose metabolism in pancreatic beta-cells. FEBS Lett. **583**(13), 2225–2230 (2009)
4. Nieto-Chaupis, H.: Nanodevices and the Internet of Bio-Nano Things for modifying insulin densities in pancreatic beta-cells through electrodynamics of Ca 2+. In: 2018 IEEE 13th Nanotechnology Materials and Devices Conference (NMDC), Portland (2018)
5. Jeon, H., et al.: Electrical force-based continuous cell lysis and sample separation techniques for development of integrated microfluidic cell analysis system: a review. Microelectron. Eng. **198**, 55–72 (2018)
6. Pizarro-Delgado, J., et al.: KCI-permeabilized pancreatic Islets: an experimental model to explore the messenger role of ATP in the mechanism of insulin production. PLOS ONE **10**(10), e0140096 (2015). https://doi.org/10.1371/journal.pone.0140096
7. Jackson, J.D.: Classical Electrodynamics, 3rd edn. Wiley, Hoboken (1999)
8. Nieto-Chaupis, H., Unluturk, B.D.: Can the nano level electrodynamics explain the very early diagnosis of diabetic nephropathy in type-2 diabetes patients? In: 2017 IEEE 17th International Conference on Nanotechnology (IEEE-NANO) (2017)
9. Arslanagic, S., Ziolkowski, R.W.: Cylindrical and spherical active coated nanoparticles as nanoantennas: active nanoparticles as nanoantennas. IEEE Antennas Propag. Mag. **59**(6), 14–29 (2017)
10. Nieto-Chaupis, H.: Evaluating the quality of service of the Internet of Bio-Nano Things in possible scenarios of applicability in the nanomedicine. In: 2017 IEEE 17th International Conference on Nanotechnology (IEEE-NANO) (2017)
11. Balanis, C.A.: Antenna Theory: Analysis and Design, 4th edn. Wiley, Hoboken (2016)
12. Piro, G., Bia, P., Boggia, G., Caratelli, D., Grieco, L.A., Mescia, L.: Terahertz electromagnetic field propagation in human tissues: a study on communication capabilities. Nano Commun. Netw. **10**, 51–59 (2016). ISSN 18787789
13. Gulbahar, B.: Theoretical analysis of magneto-inductive THz wireless communications and power transfer with multi-layer graphene nano-coils. IEEE Trans. Mol. Biol. Multi-scale Commun. **3**, 60–70 (2017). ISSN 2372-2061
14. Nafari, M., Jornet, J.M.: Modeling and performance analysis of metallic plasmonic nano-antennas for wireless optical communication in nanonetworks. Access IEEE **5**, 6389–6398 (2017). ISSN 2169-3536
15. Akyildiz, I.F., Jornet, J.M.: The Internet of Nano-Things. IEEE Wirel. Commun. **17**(6), 58–63 (2010). ISSN 1536-1284

16. Akyildiz, I.F., Pierobon, M., Balasubramaniam, S., Koucheryavy, Y.: The Internet of Bio-Nano Things. IEEE Commun. Mag. **53**(3), 32–40 (2015)
17. Jornet, J.M., Akyildiz, I.F.: The Internet of Multimedia Nano-Things in the Terahertz band. In: 18th European Wireless, Poznan, pp. 18–20, April 2012
18. Loscri, V., Vegni, A.M.: An acoustic communication technique of nanorobot swarms for nanomedicine applications. IEEE Trans. NanoBiosci. **14**(6), 598–607 (2015)
19. Nieto-Chaupis, H.: Evaluating the quality of service of the Internet of Bio-Nano Things in possible scenarios of applicability in the nanomedicine. In: 2017 IEEE 17th International Conference on Nanotechnology (IEEE-NANO) (2017)

Physics-Based Nanomedicine to Alleviate Anomalous Events in the Human Kidney

Huber Nieto-Chaupis[(✉)]

Universidad de Ciencias y Humanidades UCH, Av. Universitaria 5175,
Los Olivos, Lima39, Cercado de Lima, Peru
hnieto@uch.edu.pe

Abstract. Commonly the type-2 diabetes complications are imminent in those organs where is a substantial dependence of the microvascularity such as for instance the renal apparatus that it might be substantially affected. One of the points related to this is the degradation of the kidney functions fact that is accompanied without any symptomatology or some signals that allow the identification of the beginning of the so-called diabetes kidney disease. It is noteworthy that for large periods, clinicians have reported that type-2 diabetes patients might be potential candidates to use the dialysis machines. Therefore, to attack the problem of how to tackle down the beginning of the renal disease in type-2 diabetes would require to understand the phenomenon that is carried out in the kidney, particularly in the renal glomerulus, or glomerulus in short. In this paper we take advantage of the physics-based phenomenology to develop closed-form expressions that would describe the different scenarios by which the glomerulus is invaded by giant proteins like the albumin. Under this scenario, albumin proteins that are pushed out by the glucose dipoles in blood are expected to exert repulsion as well as attraction forces inside the glomerulus. Thus, there is a large probability as to expect that the departure of the bunches of albumin from sensitive microvascularity inside the kidneys can reach the Bowman's space and the urine formation zone. In this paper we present a study of the physics interactions inside the renal glomerulus. Essentially we use physics equations to derive the laws that govern the pass of proteins such as albumin through the layers of glomerulus. Once the physics equations are established the albumin excretion ratio is estimated. Basically, proteins do interact with glomerulus through the remaining charges along the layers and podocytes. This is crucial to determine the volume of albumin that goes to the Bowmam's space. Our study uses physics equations inside of the framework of charge electric density. The fact of measure accurately the quantity of excreted albumin with physics equations, provides capabilities to apply precise strategies in the side of the clinicians to improve the treatment in the cases where there is a potential risk to acquire the well-known diabetes kidney disease. All these methodologies encompass with the prospective Internet of Bio-Nano Things that engages an Internet network with human organs in order to anticipate any eventual abnormality or wrong functionality of organs in real-time.

Keywords: Renal damage · Kidney · Nephropathy

© ICST Institute for Computer Sciences, Social Informatics and Telecommunications Engineering 2019
Published by Springer Nature Switzerland AG 2019. All Rights Reserved
A. Compagnoni et al. (Eds.): BICT 2019, LNICST 289, pp. 14–27, 2019.
https://doi.org/10.1007/978-3-030-24202-2_2

1   Introduction

One of the main and critical complications of type-2 diabetes [1] is the so-called diabetic nephropathy by which the kidneys exhibit a degradation in their functionalities [2], fact that is manifested in the anomalous excretion of proteins. In normal scenarios, albumin and Tamm Horsfall proteins are excreted through the urine. In humans, the Tamm Horsfall protein turns out to be abundant surpassing in volume to the albumin that should be excreted to a minor extent.

In people that have had a diagnostics of type-diabetes, the concentrations of glucose in blood might not to follow a linear behavior as to the intake of carbohydrates and sugars, since the mean values of glucose might jump above the allowed ones. Therefore, during the periods where the patient exhibits large concentrations of glucose in blood, in some organs that are composed by micro microvascularity or micro veins, there is a potential risk that these organs manifest abnormalities and therefore it is established the beginning of a certain disease that would have implications in the instability of homeostasis. Due to the electric constitution, the dipoles of glucose might cancel the shield of negative charges located in the different layers of glomerulus whose task is to stop the pass of charged electrically proteins. Thus, the abundance of dipoles becomes proportional to the amount of negative charges along the layers. Once the layers are unprotected, albumin proteins pass through the urine formation zone [3]. It is sketched in Fig. 1. The rest of this paper is structured as follows: in second section we described briefly the concept of Nanomedicine used in this paper. Third section describes the physics model and the formalism to estimate the electric forces is described as well as the basis for an efficient detection of albumin proteins. Fourth section is devoted to the calculation of the repulsion force that is expected between the albumin bunching and the nano sensor. In fifth section the AER is estimated and the attained uncertainty is estimated. Finally, the conclusion of the paper is drawn.

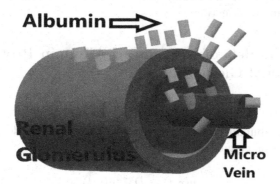

Fig. 1. Sketch of the assumed electrodynamics in the renal glomerulus: abundance of glucose might be the first cause for releasing of albumin proteins that travel through the layers of glomerulus escaping to the urine formation zone.

2   The Concept of Nanomedicine in This Paper

Since the beginning of the DKD might not have any symptomatology, classic medicine requires of test and at least a minimum window to suggest a certain pharmacology when the clinician has probed that the patient has already manifested some first symptoms that would correspond to the last phases of disease [4]. In this stage the patient might not have a back way because DKD is irreversible and a possible alteration of others organs might be observed. Certainly, the time between a first manifestation of DKD and the last phase is thought to be a decade. Therefore the idea to use the so-called Nanomedicine as an advanced methodology to alleviate and improve the renal functionality appears as an robust alternative due to:

- the emergence of nanotechnology [5],
- the apparition of prospective and advanced Internet networks [6],
- the arrival of computational systems based driven by big data,
- the wide understanding of the very beginning of DKD through physics concepts.

Figure 1 exhibits the main point of this paper: the dynamics of charges compounds in the renal glomerulus. Gian Marco Ghiggeri was the first whom observed electrical properties in human albumin [7]. This certainly opens various windows of studies in the territory entirely governed by physics laws.

Therefore the anticipation of this dynamics is inside of the territory of physics in the sense that because exists a well-defined scenario whose rules are established by physics equations, then the implementation of advanced devices that sense and perform charge measurements inside the glomerulus renal, then it might be seen as a novel methodology to estimate the level of degradation of the kidneys for instance, as well as a tool to make precise interventions against the presence of charge compounds. When these advanced devices are the so-called nanorobots, then the scenario by which the kidney is treated against next potential phases of degradation and dysfunction.

3   Physics Models of Invasion of Albumin Proteins in the Renal Glomerulus

3.1   The Diffusion Equation

The diffusion of albumin proteins through the layers, can be done through the well-known diffusion equation

$$\frac{\partial}{\partial t}u(\mathbf{r}, t) = D\boldsymbol{\nabla}^2 u(\mathbf{r}, t), \tag{1}$$

where D the diffusion's constant. As seen in Fig. 1, an acceptable coordinate system to apply Eq. (1) becomes the cylindrical system by which one can anticipate that the albumin would exhibit radial trajectories. Under this reference system, then we can write down:

$$\frac{1}{D}\frac{\partial}{\partial t}u(\mathbf{r},t) = \frac{1}{r}\frac{\partial}{\partial r}\left(r\frac{\partial}{\partial r}u(\mathbf{r},t)\right) + \left(\frac{1}{r^2}\frac{\partial^2}{\partial\theta^2} + \frac{\partial^2}{\partial z^2}\right)u(\mathbf{r},t). \tag{2}$$

The usage of the method of separation of variables lead us to write the function $u(\mathbf{r},t) = R(r)\Theta(\theta)Z(z)T(t)$, so that we can obtain

$$\frac{1}{DT(t)}\frac{dT(t)}{dt} = -\lambda \tag{3}$$

$$\frac{1}{rR(r)}\frac{\partial}{\partial r}\left(r\frac{\partial R}{\partial r}\right) + \left(\frac{1}{\Theta r^2}\frac{\partial^2\Theta}{\partial\theta^2} + \frac{1}{Z}\frac{\partial^2 Z}{\partial z^2}\right) = -\lambda. \tag{4}$$

Eq. (2) is solved in a straightforward manner resulting in $T(t) = T_0\mathrm{Exp}(-\lambda Dt)$. However, Eq. (3) implies that $\frac{1}{Z}\frac{\partial^2 Z}{\partial z^2} = -m^2$ so that $\frac{1}{rR(r)}\frac{\partial}{\partial r}\left(r\frac{\partial R}{\partial r}\right) + \frac{1}{\Theta r^2}\frac{\partial^2\Theta}{\partial\theta^2} + k^2$ and $\frac{1}{\Theta}\frac{\partial^2\Theta}{\partial\theta^2} = -\ell^2$ with ℓ an integer number, and finally we have

$$\phi^2\frac{d^2 R}{d\phi^2} + \phi\frac{dR}{d\phi} + \left[\phi^2 - \ell^2\right]R(r) = 0, \tag{5}$$

with $\phi = qr$, and $q = \sqrt{\lambda - m^2}$ and which is essentially the Bessel's equation with general solution: $R(\phi) = \mathcal{A}J_\ell(\phi) + \mathcal{B}J_{\ell+1}(\phi)$. On the other hand, $\Theta(\theta) = \mathcal{C}_\ell\mathrm{Sin}(\ell\theta) + \mathcal{D}_\ell\mathrm{Cos}(\ell\theta)$ and $Z(z) = \mathcal{A}_m\mathrm{Sin}(mz) + \mathcal{B}_m\mathrm{Cos}(mz)$. With the boundary conditions, $m = \frac{n_1\pi}{z_1}$ and $\ell = \frac{n_2\pi}{\theta_1}$, with $n_{1,2}$ integer numbers, and z_1 and θ_1 values corresponding to the geometry of glomerulus and $q \approx \sqrt{\lambda}$,

Boundary Conditions. According to Fig. 1 we have imposed two well-defined boundary conditions for the radial part:

- $R(qr_1) = 0$ for $t_0 = 0$ that means that the flux of albumin proteins starts from this radius $r = r_1$, whereas for $r = r_2$ exists a substantial flux of bunches of albumin and other types of proteins that are crossing the glomerulus [8] whose length might be given by the distance $r_2 - r_1 \approx 100$ nm.
- And $R(qr_2) = \mathcal{R}$ that denotes that the concentration of proteins just in $r = r_2$ has a determined values namely $\mathcal{R}$ for $t > t_0$. For this time the bunches of proteins have evolved in time. The fact of imposing a solution with the index $\ell + 1$ is due to that these boundary conditions a solution of the form $AJ_\ell + BJ_{\ell+1}$ is not admitted, because it results in a trivial solution.

Thus one would expect that a nano sensor is located near to the region of proteins evacuation to guarantee that negatively charged proteins might be enough efficient to discriminate them from the noise. By knowing the size of the albumin protein of order of 10 nm a possible size of the nano sensor would be of order of 100 nm. In this manner a closed-form solution of (1) is written as

$$\rho_A(\mathbf{r},t) = \rho_0 \mathrm{Exp}^{[-\lambda D(t-t_0)]} \times$$

$$\times \sum_{\ell=-\infty}^{\infty} \frac{\mathrm{Sin}\left(\frac{n_1\pi z}{z_1}\right)\mathrm{Sin}\left(\frac{n_2\pi\theta}{\theta_1}\right)[J_{\ell+1}(qr_1)J_\ell(qr) - J_{\ell+1}(qr)J_\ell(qr_1)]}{J_\ell(qr_2)J_{\ell+1}(qr_1) - J_\ell(qr_1)J_{\ell+1}(qr_2)}, \tag{6}$$

Thus the total charge derived from this solution is $Q_A = \int \rho_A(\mathbf{r},t)dV$.

3.2   The Jackson Equation

Again, as seen in Fig. 1, the assumption that the glomerulus follows a cylindrical geometry makes us to test a different approach entirely based on classical electrodynamics. In contrast to the previous approach classical electrodynamics [9] does not contain the time so that all solutions are fully independent of time. **This view fits in a convenient manner the modeling of a charged nano sensor deployed inside the layers of kidney being one of them the renal glomerulus.**

Therefore the nano sensor is modeled through a charge Q inside of a volume belonging to the renal glomerulus. Thus this charge produces a electric potential to a distance measured by the vector $\mathbf{R} = \mathbf{r} - \mathbf{r}'$, where the full charge Q is located in $\mathbf{r}$ and the potential is calculated in $\mathbf{r}'$. The solution to this problem turns to be as the so-called Jackson potential. The potential is given in [11], $\Phi(\mathbf{r},\mathbf{r}')$ depends is the cylindric parameters L and a and the full charge containing inside Q. Because the charge Q is fixed there is not time dependence as argued claimed above. According to the Poisson's equation the associated charge density satisfies the differential equation given by:

$$\nabla^2\Phi(\mathbf{r},\mathbf{r}') = -\frac{\rho(\mathbf{r},\mathbf{r}')}{\epsilon}, \tag{7}$$

The full closed-form solution is expressed as:

$$\rho_B(\mathbf{r},\mathbf{r}') = -\epsilon\nabla^2\Phi(\mathbf{r},\mathbf{r}') =$$

$$-\frac{2Q}{\pi L a^2}\sum_{\ell,m,n} \frac{\mathrm{Exp}[im(\theta-\theta')]\mathrm{Sin}\left(\frac{\ell\pi z}{L}\right)\mathrm{Sin}\left(\frac{\ell\pi z'}{L}\right)}{\left[\left(\frac{\xi_{m,n}}{a}\right)^2 + \left(\frac{\ell\pi}{L}\right)^2\right]J_{m+1}^2(\xi_{m,n})}\frac{\ell\pi}{4L}J_{m+1}\left(\frac{\xi_{m,n}r'}{a}\right)\times$$

$$\frac{\xi_{m,n}}{4a^2}[\xi_{m,n}J_{-2+m}\left(\frac{\xi_{m,n}r}{a}\right) - J_m\left(\frac{\xi_{m,n}r}{a}\right) - \xi_{m,n}J_m\left(\frac{\xi_{m,n}r'}{a}\right) - J_{2+m}\left(\frac{\xi_{m,n}r}{a}\right)], \tag{8}$$

with a sign negative, therefore the total charge can be written as

$$Q = -\int \epsilon\nabla^2\Phi(\mathbf{r},\mathbf{r}')d^3r' \tag{9}$$

where Q that is located in r', θ', z' denotes the charge of the nano device that is the source of electric field in the point r, θ, z.

4 Closed-Form Expressions of the Electric Charges and Force

With Eqs. 6 and 9 the charges that under interaction will produce a electric force, the full electric force created by the repulsion of the nano sensor and the negative charged bunches of albumin is obtained in a straightforward manner from

$$\mathcal{F} = \int \frac{QQ_A d^3 \mathbf{r}'}{|\mathbf{r} - \mathbf{r}'|} \tag{10}$$

with the previous equations the full electric force that would involve the nano sensor and the bunching of albumin reads

$$|\mathcal{F}| = \frac{1}{4\pi\epsilon} \int \frac{\int \epsilon \nabla^2 \Phi(\mathbf{r}, \mathbf{r}') d^3 \mathbf{r}' \int \rho(\mathbf{r}, t) d^3 \mathbf{r}}{|\mathbf{r} - \mathbf{r}'|^2} \tag{11}$$

that is actually the strength of the electric force in the radial direction in the sense that the albumin escapes along the radial direction under the assumption that the glomerulus has the cylindrical shape. Due to the fact that the bunches of albumin are charged electrically compounds, then is of interest to calculate the full electric charge that might be sensed by a nano sensor.

Fixed Volume of Nano Sensor
In order to extract a value of this electric force, we assume that the charge of nano sensor follows $Q = \rho \Delta V = \rho \Delta z R \Delta R \Delta \theta$ thus in this manner we have

$$|\mathcal{F}_R| = \frac{1}{4\pi} \int \frac{\int \rho(\mathbf{r}, t) d^3 \mathbf{r} \rho \Delta z R \Delta R \Delta \theta}{|\mathbf{R}|^2} \tag{12}$$

while the nano device dimension left fixed $\Delta z R \Delta R \Delta \theta = $ constant then the volume is compact,

$$|\mathcal{F}_R| = \frac{1}{4\pi} \int \frac{\int \rho(\mathbf{r}, t) d^3 \mathbf{r} Q_N}{|\mathbf{R}|^2} \tag{13}$$

$$|\mathcal{F}_R| = \frac{1}{4\pi} \int J_{\ell+1}(qr_1) J_\ell(qr) \frac{Q_N}{r^2} dV = \frac{1}{4\pi} Q_N \int^r \frac{J_{\ell+1}(qr_1) J_\ell(qr)}{r^2} dV$$

by taking into account the factor 2 in Eq. 8 we can arrive to:

$$|\mathcal{F}_R| = \frac{1}{4\pi} 2 Q_N d\theta dz \int^r \frac{J_{\ell+1}(qr_1) J_\ell(qr)}{r^2} dr \tag{14}$$

with the integration over the product of two integer-order Bessel functions that can be done in a closed-form yielding:

$$|\mathcal{F}_R| = 2 Q_N \Delta\theta \Delta z \left[\frac{J_n(r) J_m(r)}{r(m+n+1)} + \frac{J_{n-1}(r) J_m(r) + J_n(r) J_{m-1}}{(m+n)^2 - 1} \right] \tag{15}$$

In Fig. 2 are shown the electric forces calculated with Eq. (15). The fact of the assumption that the nano sensor has a well-defined volume and charge the calculation as seen in (15) has shown to be exact. In Fig. 2 the different manifestations of the electric force have been plotted. For example in Top panel the force is plotted but only the first term of numerator is taken. The cases (a) to (e) denote the orders m and n. The peaks here are understood as all those distances by which the flux of albumin is maximum. The position of these peaks follows the used orders in the Bessel functions. In middle plot the normalized force taking into account both terms of numerator is plotted. For example the case (A) displays a morphology that can be understood in terms of the sign of the charge distributions. A first peak is seen in 0.20 µm and subsequently a minor peak is seen in 1.23 µm. Although this minor peak might be explained in terms of the decreasing of flux of albumin [10], also it can be interpreted as the flipping of the sign of the nano sensor. The usage of high orders of the Bessel functions displays similar shapes as (A). The cases (B), (C), and (D) emerge as the fact that in large distance from the micro microvascularity the intensity falls down since the possible apparition of positive charges would cancel a substantial part of the bunches of albumin so that the force intensity becomes small. The black arrows point the peaks by which one expects that the flux is substantial as for limiting the strength of the electric force. In bottom panel the formation of successive peaks is seen. Here we consider the whole expression Eq. 15. Interestingly the curve (B) for the distances between 1.5 µm and 3.0 µm the electric force turned out to be negative with the minor peaks in (B) and (C) still with positive values. In large distances, curve (A) recover the initial sign and becomes positive. Clearly this behavior is responding to one of character entirely oscillatory more than a linear dynamics as assumed previously. In virtue to this we can establish the following:

- Oscillations of the sign of the electric force might have as origin the unbalancing of the charge distributions along the renal glomerulus,
- The nano sensor has as function to change its sign through external bio cyber interfaces,
- The pass of large bunches of albumin might to trigger nonlinear dynamics in the electrical interactions.

4.1   The Full Electric Force

A special scenario constitutes the one where the electric force is dependent on the time. To include the time in the full formulation of the strength of the electric force we shall use the solution of the diffusion equation that involve the exponential dependent on the time. Considering the existence of attraction and repulsion forces, then we can write down

$$F_R(r,t) = \frac{2\mathcal{Q}_N \rho_0 \mathrm{Exp}(-\lambda Dt)}{\pi \epsilon La^2} \sum_{\ell,m,n} \int 4\pi r^2 dr J_{-2+m}\left(\frac{\xi_{m,n} r}{a}\right) J_{\ell-1}(qr), \quad (16)$$

$$F_A(r,t) = -\frac{2\mathcal{Q}_N \rho_0 \mathrm{Exp}(-\lambda Dt)}{\pi \epsilon La^2} \sum_{\ell,m,n} \int 4\pi r^2 dr J_{2+m}\left(\frac{\xi_{m,n} r}{a}\right) J_{\ell+1}(qr), \quad (17)$$

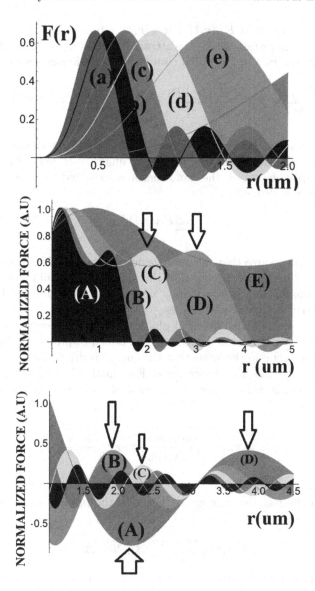

Fig. 2. Top: The electric force (expressed in arbitrary units versus the radial distance measured in μm. Middle and bottom: the normalized electric force as function of the radial distance expressed in μm. The order of the curves follows the order of the Bessel function from the lower until the higher order. In these simulations, (A)–(E) run between from the 0-order to the 7-th order.

by which is imminent that $F_R(r,t) + F_A(r,t) = 0$.

To investigate in certain details the attraction forces inside the renal glomerulus we consider the rate of full radial component electric force is written as:

$$\frac{\Delta F_A(r,t)}{\Delta r} = -\frac{R^2 \mathrm{Exp}(-\lambda Dt)}{(1-\gamma)^2} \sum_{\ell,m,n} \frac{J_{2+m}(\frac{\xi_{m,n}r}{a})J_{\ell+1}(qr)}{r^2}. \tag{18}$$

Equation (18) is relate out with the energy that would expend the nano device once that it acquires mechanical motion. The quantity R denotes the radius of the spherical space by which the nano sensor is expected to be moving inside. Thus this energy is defined as

$$\mathcal{E} = \frac{||\Delta F_A(r,t)||}{\Delta r} \mathcal{A} \tag{19}$$

where $\mathcal{A}$ the effective area that the nano device passes on it. In terms of prospective and realistic usage is expected that the nano device should move inside the smallest portions of area in the kidney in order to avoid damage by physical contacts between the device and tissues.

Since we are dealing with classical physics, the entire energy due to electric interactions between the nano device and the bunch of proteins is perceived as one of harmonic origin in the sense that the energy is driven by $\mathcal{E} = m\omega^2 x^2/2 + m_N v^2/2$ where m_D the nanodevice mass. For small velocities the pure kinetic term is neglected, in this manner the period of oscillation of the nanodevice appears to be as:

$$T = \pi r \sqrt{\frac{2m_N \Delta r}{||\Delta F(r,t)||\mathcal{A}}} \approx \pi r \sqrt{\frac{2m_N}{\mathcal{E}}}. \tag{20}$$

The Physical Action

Starting from the fact that the action becomes the product of energy by time then with (19) and (20) we have

$$S = \pi r \sqrt{\mathcal{A}||\Delta F(r,t)||2m_N \Delta r} \tag{21}$$

that implies that the action is proportional to the root square of the mass of the nano sensor, denoting the importance of the physics properties of the nano sensor to adjust electric dynamics in the renal glomerulus. This fact is seen also from the angle of the powering engineering that targets to provide energy to the nano sensor in order to extend the most large lifetimes when is working in the renal glomerulus.

In order to numerically evaluate Eq. 18 we have plotted in Fig. 3 distributions of dF/dr (assuming SI units) for the cases of repulsion (same sign) and attraction (opposite signs) between the bunch of albumin and the nano sensor. We can see that in both cases that the forces fall down as the increasing of the order of the Bessel function as seen in Fig. 2.

Certainly the large peaks is done for the first orders. Roughly speaking the flipping of the sign in the force is driven by the net amount of positively charged proteins. Physically speaking this makes that the proteins are attracted to the proximity of the nano sensor. Contrarily, the case where the bunch and nano sensor are same sign the **repulsion can be seen a mechanism that expels and detains the exit of successive trains of bunches of albumin through the Bowman's space.**

In Table 1 are listed the input values that yields a crude estimate of the period of oscillation of the nano sensor. Thus under the assumption that 1 pulse electromagnetic is emitted by the nano sensor per an entire period of 0.04 s under the event of attraction or repulsion then a total of 90,000 pulses per hour are expected to be sent by the nano sensor to the bio cyber interface inside the framework of the Internet of Bio-Nano Things. Clearly the expected bio cyber device is expected to be able to discriminate the true signal *i.e* the albumin proteins against noise or background such as the Tamm-Horsfall proteins. Therefore this prospective nano sensor is expected that might learn from the experienced oscillations to recognize albumin proteins from others proteins that would not turn out to be hostile chemical compounds against the glomerular zone as well as the intrinsic functionalities of kidney [11].

Table 1. Numerical Estimates of Eq. 20.

Quantity	Value	Error
$\mathcal{E}$	10^{-10} Nm	±1%
m_N	10^{-6} Kg	±0.5%
r	10^{-4} m	±0.05%
T	$4.42 \ 10^{-2}$ s	±0.75 %

5 Estimation of the Albumin Excretion Rate

5.1 Theoretical Derivation

Normally, the AER parameter used extensively in the diagnosis of the early renal disease is expressed in terms of mg/dL as well as mg/24 h. Advanced phases of the so-called albuminuria can reach or surpass 300 mg/24 h. Thus, from Eq. 20 the energetic observable is related to the kinetic energy of the bunching of albumin leaving the renal glomerulus,

$$\mathcal{E} = \frac{M_A}{2} v_A^2 \tag{22}$$

when the bunching performs the spatial displacement is any direction then we have

$$\mathcal{E} = \frac{M_A}{2} \frac{\ell^2}{T^2}, \tag{23}$$

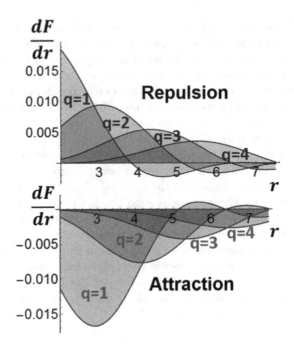

Fig. 3. The rate of change of electric force per radial distance for both scenarios: repulsion (up) and attraction (down). The plots are done up to the fourth order of the Bessel functions.

we focus now on the quantity T that is also seen as the time or period by which the bunching is displaced a length ℓ. Thus in a first instance the AER as function of the mass of albumin per unit of time can be expressed as

$$\text{AER} = \frac{2\mathcal{E}T}{\ell^2} \tag{24}$$

inserting Eq. 20 then AER is read as

$$\text{AER} = \frac{2\mathcal{E}}{\ell^2}\pi r\sqrt{\frac{2m_N\Delta r}{||\Delta\boldsymbol{F}(r,t)||\mathcal{A}}} \tag{25}$$

and the $\mathcal{C} = 2\pi r$ the circumference that defines the area occupied by the pass of the bunching of albumin through the glomerulus, then one gets that

$$\text{AER} = \frac{\mathcal{C}}{\ell^2\Delta r}\sqrt{2m_N\Delta r||\Delta\boldsymbol{F}(r,t)||\mathcal{A}} \tag{26}$$

that constitutes the theoretical equation of AER in humans due to the anomalous transit of albumin proteins through the kidney targeting the urine formation zone.

5.2 Numerical Approximations

The peaks of the electric force as shown in Fig. 2 (middle) yields the value: 0.7 for $r = 2.0\,\mu m$ and $3.0\,\mu m$ for instance. It is interpreted as the maximum value of proteins excretion. Now we pass to estimate the AER from these values. For this end, we use the expression

$$AER = \frac{M_{PR}}{v_{PR} \times s} \times V_{TOT} \times Exp\left[-\left(\frac{I - I_0}{\Delta I_0} \right)^2 \right] \tag{27}$$

where M_{PR} mass of protein of albumin, V_{TOT} total volume for both kidneys, and v_{PR} full volume of proteins. We have introduced the Gaussian profile $Exp\left[-\left(\frac{I-I_0}{\Delta I_0}\right)^2\right.$ where is assumed that the AER value is depending on the capability of the nano device to get a precise value of the intensity of the electric force I_0. Since one expects that the nano device will be sending in average 90 K pulses/h then the error is set in a first instance to 10^{-4} Firstly, we assume that the net number of human glomerulus is 10^6 (both kidneys). Thus, we can estimate $V_{TOT} = 0.7\ 64\ (10\,\mu m)^3\ 10^6$, where $0.7 = p_\ell$ denotes the estimated of the peak of electric force as seen in Fig. 2 and ℓ the order of the Bessel function. On the other hand, $v_{PR} = \pi(2\,nm)^2\ 2\,nm = 8\pi\ (nm)^3$ (assuming a cylindrical geometry). Finally, $M_{ALB} = 1{,}6\ (1\,nm)^3$ kg, which is the albumin's mass. A straightforward calculation, yields AER $\approx 2.22\ (1 \times 10^{-9})$ kg/s, where $1n = 10^{-9}$. This value is actually of order of 420 mg/day being greater that the setpoint established by the nephrologist of order of 300 mg/day. This clearly is perceived in somewhat as the very beginning of the DKD in patients having an older diagnosis of type-2 diabetes of order of 10 years in average and in conjunction with a poor self-care that might be a crucial cause of the beginning of the DKD.

5.3 The Error of AER Calculation

Because the AER is a composition of various terms being each one of different nature in the sense that a precise value might involve a certain uncertainty, so a full equation that enclose all possible sources of errors can be written as

$$\frac{\Delta AER}{AER} = \sqrt{\left(\frac{\Delta p_\ell}{p_\ell} \right)^2 + \left(\frac{\Delta V_{TOT}}{V_{TOT}} \right)^2 + \left(\frac{\Delta I_0}{I_0} \right)^2 + \left(\frac{\Delta M_{ALB}}{M_{ALB}} \right)^2} \tag{28}$$

where the main source of error would come from the capacity of the model to make predictions of peak of the electric force sensed by the nano device. However the error can also be subject to systematic errors by the assumptions of the theoretical model. A rapid calculation of ΔAER yields 6%, roughly.

Figure 4 displays various scenarios where the intensity of the electric force as given by Eq. 15 are plotted. We can see the existence of a white and yellow area denoting the signal as seen by the nano sensor. We use the technique of Smooth-Density-Histogram and bandwidth methodology: top, middle and bottom: 0.10, 0.15 and 0.3 respectively. Top middle and bottom left plots would display what we can be sensed by a nano sensor.

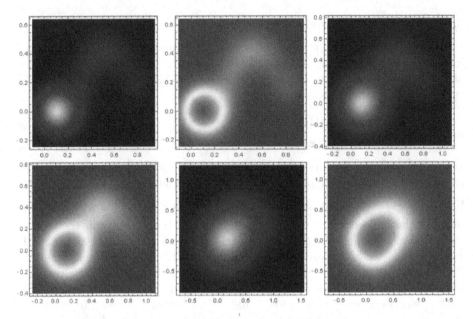

Fig. 4. Smooth-Density-Histograms for different scenarios of the intensity of electric force by using Eq. 10. Plots were done with the Smooth-Density-Histogram and bandwidth methodology: top, middle and bottom: 0.10, 0.15 and 0.3 respectively. The centers of plots denote the spatial location of the signal: albumin protein whereas their tails are associated to noise or another type of protein. The expected nano sensor would register data solely from the centered concentration as a first sign of a possible anomalous event in the renal apparatus. (Color figure online)

6   Conclusion

In this paper we have developed a theory that would estimate the AER a parameter of importance to make a accurate diagnosis of the renal disease in those type-2 diabetes patients. We turned to the territory of the Nanomedicine in the sense that we have proposed a technique that would identify the first phases of diseases as well as a method of surveillance through nano sensors has been explained. Since albumin are negatively charged proteins, it open the possibility to detect them with electric interactions governed by physics laws. Numerical estimations have yielded a crude estimate of 90 K pulses per hour being enough statistics to perform analysis of signal and noise acceptation and rejection, fact the one would help to nephrologist to reconfigure the treatment. The compound is understood to be a bunch of proteins of albumin. These giant proteins are leaving the glomerulus because their electrodynamics with the shielding of charges over the inner and outer layers of glomerulus. When the maximum value of the density of charge or compound is estimated, it enters in a straightforward estimation of AER, yielding a value of 450 mg/day. According to the clinical tests, this value belongs to the case of a type-2 patient showing the very beginning of the DKD. These results would support the idea that the deployment of a nano sensor near to the glomerulus might be advantageous for the anticipation of the very beginning of DKD, by assuming the central hypothesis: dynamics of the bunches of albumin is entirely governed for repulsion and attraction electric forces.

References

1. World Health Organization: Definition, diagnosis and classification of diabetes mellitus and its complications. Part 1: diagnosis and classification of diabetes mellitus report of a WHO consultation, Geneva, Switzerland (1999)
2. Shankland, S.J., Pollak, M.R.: A suPAR circulating factor causes kidney disease. Nat. Med. **17**, 926–927 (2011). https://doi.org/10.1038/nm.2443
3. Reiser, J.: Akt2 relaxes podocytes in chronic kidney disease. Nat. Med. **19**, 1212–1213 (2013). https://doi.org/10.1038/nm.3357
4. Strain, W.D.: Albumin excretion rate and cardiovascular risk could the association be explained by early microvascular dysfunction? Diabetes **54**, 1816–1822 (2005)
5. Nieto-Chaupis, H.: Closed-form solutions of the diffusion equation to model prospective nanodevice to anticipate diabetes kidney disease through electric forces. In: 2018 IEEE 13th Nanotechnology Materials and Devices Conference (NMDC), Portland, OR, USA (2018)
6. Nieto-Chaupis, H.: Prospects for anticipating kidney damage in type-2 diabetes patients through the sensing of albumin passing through the renal glomerulus. In: 2017 IEEE EMBS International Conference on Biomedical and Health Informatics (BHI), Orlando, FL, USA, 16–19 February 2017
7. Ghiggeri, G.M., Candiano, G., Delfino, G., Queirolo, C.: Electrical charge of serum and urinary albumin in normal and diabetic humans. Kidney Int. **28**, 168–177 (1985)
8. Tonneijck, L., et al.: Glomerular hyperfiltration in diabetes: mechanisms, clinical significance, and treatment. J. Am. Soc. Nephrol. JASN **28**, 1023–1039 (2017)
9. Jackson, J.D.: Classical Electrodynamics, 3rd edn. Wiley, Hoboken (1999)
10. Sun, Y.B.Y., Qu, X., Zhang, X., Caruana, G., Bertram, J.F., Li, J.: Glomerular endothelial cell injury and damage precedes that of podocytes in adriamycin-induced nephropathy. PLoS ONE **8**(1), e55027 (2013). https://doi.org/10.1371/journal.pone.0055027
11. Assady, S., Wanner, N., Skorecki, K.L., Huber, T.B.: New insights into podocyte biology in glomerular health and disease. J. Am. Soc. Nephrol. JASN **28**, 1707–1715 (2017). https://doi.org/10.1681/ASN.2017010027

Bio-inspired System Identification Attacks in Noisy Networked Control Systems

Alan Oliveira de Sá[1,2(✉)], António Casimiro[3],
Raphael Carlos Santos Machado[4,5], and Luiz Fernando Rust da Costa Carmo[2,4]

[1] Admiral Wandenkolk Instruction Center, Brazilian Navy,
Brasilia, RJ, Brazil
alan.oliveira.sa@gmail.com
[2] Institute of Mathematics/NCE, Federal University of Rio de Janeiro,
Rio de Janeiro, RJ, Brazil
[3] Department of Informatics, Faculty of Sciences of the University of Lisboa,
Lisbon, Portugal
casim@ciencias.ulisboa.pt
[4] National Institute of Metrology, Quality and Technology,
Rio de Janeiro, RJ, Brazil
{rcmachado,lfrust}@inmetro.gov.br
[5] Rio de Janeiro Federal Center for Technological Education,
Rio de Janeiro, RJ, Brazil

Abstract. The possibility of cyberattacks in Networked Control Systems (NCS), along with the growing use of networked controllers in industry and critical infrastructures, is motivating studies about the cybersecurity of these systems. The literature on cybersecurity of NCSs indicates that accurate and covert model-based attacks require high level of knowledge about the models of the attacked system. In this sense, recent works recognize that Bio-inspired System Identification (BiSI) attacks can be considered an effective tool to provide the attacker with the required system models. However, while BiSI attacks have obtained sufficiently accurate models to support the design of model-based attacks, they have demonstrated loss of accuracy in the presence of noisy signals. In this work, a noise processing technique is proposed to improve the accuracy of BiSI attacks in noisy NCSs. The technique is implemented along with a bio-inspired metaheuristic that was previously used in other BiSI attacks: the Backtracking Search Optimization Algorithm (BSA). The results indicate that, with the proposed approach, the accuracy of the estimated models improves. With the proposed noise processing technique, the attacker is able to obtain the model of an NCS by exploiting the noise as a useful information, instead of having it as a negative factor for the performance of the identification process.

© ICST Institute for Computer Sciences, Social Informatics and Telecommunications Engineering 2019
Published by Springer Nature Switzerland AG 2019. All Rights Reserved
A. Compagnoni et al. (Eds.): BICT 2019, LNICST 289, pp. 28–38, 2019.
https://doi.org/10.1007/978-3-030-24202-2_3

Keywords: Security · Networked Control Systems ·
Cyber-physical systems · System identification ·
Backtracking Search Algorithm · Bio-inspired algorithm

1 Introduction

The use of communication networks to integrate controllers and physical processes in a Networked Control Systems (NCS), such as shown in Fig. 1, aims to improve management and operational capabilities, as well as reduce costs [10]. However, this integration also exposes the physical plants to new threats originated in the cyber domain.

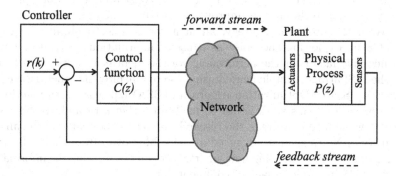

Fig. 1. Networked control system.

The possibility of sophisticated and large impact attacks in Networked Control Systems (NCS) became unprecedentedly concrete after the launch of the Stuxnet worm [6]. The example of such cyber-physical attack – which is not unique –, along with the growing use of networked controllers in industry and critical infrastructures, has been motivating studies about the cybersecurity of NCSs. In this context, there is a research effort to characterize vulnerabilities, understand attack strategies, and propose security solutions for NCS [1,3,7–14].

The literature on cybersecurity of NCSs [1,9–12,14] indicates that accurate and covert offensives require high level of knowledge about the models of the attacked system. Examples of covert attacks that agree with this statement are provided in [11,12]. In these works the attacks are performed by a man-in-the-middle (MitM), where the attacker needs to know the model of the attacked plant to covertly manipulate the system by injecting false data in both forward and feedback streams. The covertness of the attacks shown in [11,12] is analyzed from the perspective of the signals arriving to the controller, and depends on the difference between the actual model of the plant and the model known by the attacker. In [1], the authors demonstrate another covert offensive where the attacker, aware of the system's model, injects an attack signal in the NCS to steal water from a canal system located in Southern France.

However, in [1,11,12,14], where the attacks intrinsically require knowledge about the NCS models, it is not described how such knowledge is obtained by the attacker. It is just stated that a model is previously known to subsidize the design of those attacks. More recently, in [9,10], the authors propose two Bio-inspired System Identification (BiSI) attacks to fill this gap. They demonstrate how the data required to design Denial-of-Service (DoS) or Service Degradation (SD) attacks may be obtained using bio-inspired metaheuristics. Specifically, the attacks proposed in [9,10] are used to obtain the linear time-invariant (LTI) transfer functions of NCS devices – be it a controller [10], a plant [10], or both in a open loop transfer function [9].

While BiSI attacks have obtained sufficiently accurate models to support the design of model-based attacks, they have demonstrated loss of accuracy in the presence of noisy signals [9]. To overcome this constraint, this work proposes a noise processing technique to improve the accuracy of BiSI attacks in noisy NCSs. With the proposed strategy, an attacker is able to obtain the model of an NCS by exploiting the noise as a useful information, instead of having it as a negative factor for the performance of the identification process. In this paper, the BiSI attack is implemented using the bio-inspired metaheuristic called Backtracking Search Optimization Algorithm (BSA) [2]. It is worth mentioning that the purpose of this work is not to facilitate cyber-attacks in NCSs. With this study, we aim to encourage the research for techniques capable to enhance the security of NCSs against advanced attacks. Moreover, from the NCS owner perspective, it is worth knowing how an attacker can obtain valuable information about the NCS in case of a lack of confidentiality.

The next sections of this work are organized as follows. Section 2 provides a brief description about the BSA. Section 3 explains the novel noise processing strategy for BiSI attacks. Section 4 shows the results obtained when the noise processing strategy herein proposed is used to support a BiSI attack. Finally, Sect. 5 brings the conclusions of this work.

2   Backtracking Search Algorithm

This section describes the basic concepts of the BSA, in order to provide a clear understanding about the algorithm parameters that are adjusted when implementing a BSA-based BiSI attack. The BSA is a bio-inspired metaheuristic that searches for solutions of optimization problems using the information obtained by past generations [2] – or iterations. According to [2], its search process is metaphorically analogous to the behavior of a social group of animals that, at random intervals returns to hunting areas previously visited for food foraging. The general, evolutionary like, concept of the BSA is shown in Algorithm 1.

Algorithm 1. BSA

> **begin**
>> Initialization;
>> **repeat**
>>> Selection-I;
>>> **Generate new population**
>>>> Mutation;
>>>> Crossover;
>>>
>>> **end**
>>> Selection-II;
>>
>> **until** *Stopping Condition*;
>
> **end**

At the Initialization stage, the algorithm generates and evaluates the initial population $\mathcal{P}_0$ and sets the historical population $\mathcal{P}_{hist}$. The latter acts as the memory of the BSA.

In the first selection stage (Selection-I), the algorithm randomly determines, based on an uniform distribution U, whether the current population $\mathcal{P}$ should be kept as the new historical population and, thus, replace $\mathcal{P}_{hist}$ (*i.e.* if $a <$ $b \mid a, b \sim U(0, 1)$, then $P_{hist} = P$). After that, it shuffles the individuals of $\mathcal{P}_{hist}$.

The mutation operator creates $\mathcal{P}_{mod}$, which is the preliminary version of the new population $\mathcal{P}_{new}$. The computation of $\mathcal{P}_{mod}$ is performed according to (1):

$$\mathcal{P}_{mod} = \mathcal{P} + \eta \cdot \Gamma(\mathcal{P}_{hist} - \mathcal{P}), \tag{1}$$

wherein η is empirically adjusted through simulations and $\Gamma \sim N(0, 1)$, with N being a normal standard distribution. Thus, $\mathcal{P}_{mod}$ is the result of the movement of $\mathcal{P}$'s individuals in the directions established by vector $(\mathcal{P}_{hist} - \mathcal{P})$. In order to create the final version of $\mathcal{P}_{new}$, the crossover operator randomly combines individuals from $\mathcal{P}_{mod}$ and $\mathcal{P}$, also following a uniform distribution.

In the second selection stage (Selection-II), the algorithm evaluates the elements of $\mathcal{P}_{new}$ using a fitness function f, selects the elements of $\mathcal{P}_{new}$ with better fitness than the ones in $\mathcal{P}$, and replaces them in $\mathcal{P}$. Hence, $\mathcal{P}$ includes only new individuals that have evolved. The algorithm iterates until the stopping condition is met. When it occurs, the BSA returns the best solution found.

Note that the algorithm has two parameters that are empirically adjusted: the size $|\mathcal{P}|$ of its population $\mathcal{P}$; and η, that establishes the amplitude of the movements of the individuals of $\mathcal{P}$. The parameter η must be adjusted to assign to the algorithm both good exploration and exploitation capabilities. With this parameters set, the BSA is used to search for the global minimum of the fitness function f described in Sect. 3.

3   Noise Processing Technique for BiSI Attacks

The purpose of the technique presented in this section is to use the white gaussian noise that may be present in an NCS – such as in [9] – in favor of a BiSI attack. With this technique, an attacker is able to accurately estimate the models of an NCS by exploiting the noise as a useful information, instead of having it as a negative factor for the performance of the identification process – which happened in previous implementations of BiSI attacks [9].

The first step of the attack is to eavesdrop the input $i(k)$ and output $o(k)$ signals of the device to be identified, represented in Fig. 2. The device can be a controller or a plant. The signals are captured during a monitoring period containing T samples.

Fig. 2. Device to be identified.

After that, the attacker selects every sample of the eavesdropped input signal $i(k)$ that exceeds a predefined threshold Ω, *i.e.* if (2) is satisfied:

$$i(k) > \Omega, \tag{2}$$

Each sample selected from $i(k)$ according to (2) is referred to as i_n, wherein $n \in \mathbb{Z}_+^*$ is a sequential index number for each selected sample, as exemplified in Fig. 3. Additionally, every time that (2) is satisfied, the attacker also stores a portion $o_n(k)$ of the output signal $o(k)$. As represented in Fig. 3, each portion $o_n(k)$ selected from $o(k)$ starts when its respective i_n occurs. Each portion $o_n(k)$ encompasses a sequence of τ samples.

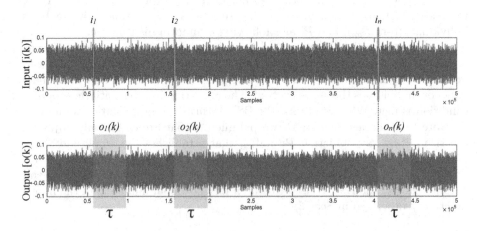

Fig. 3. Selection of noise portions.

After selecting all i_n and $o_n(k)$ from the eavesdropped signals, the attacker computes $\mathcal{I}$ and $\mathcal{O}(k)$ according to (3) and (4), respectively:

$$\mathcal{I} = \frac{\sum\limits_{n=1}^{\mathcal{N}} i_n}{\mathcal{N}}, \tag{3}$$

$$\mathcal{O}(k) = \frac{\sum\limits_{n=1}^{\mathcal{N}} o_n(k)}{\mathcal{N}}, \tag{4}$$

wherein $\mathcal{N}$ is the index number of the last sample i_n obtained from $i(k)$ based on (2). In the present approach, $\mathcal{I}$ corresponds to the amplitude of an impulse signal $\mathcal{I}(k)$ (5) that, when applied to $G(z)$, produces $\mathcal{O}(k)$ – the impulse response function of $G(z)$.

$$\mathcal{I}(k) = \mathcal{I}\delta(k). \tag{5}$$

Now, to estimate $G(z)$, the attacker applies $\mathcal{I}(k)$ to the input of an estimated model $G_e(z)$ defined by (6):

$$G_e(z) = \frac{\mathcal{Z}[\hat{O}(k)]}{\mathcal{Z}[\mathcal{I}(k)]} = \frac{\alpha_p z^p + \alpha_{p-1} z^{p-1} + \dots + \alpha_1 z^1 + \alpha_0}{z^q + \beta_{q-1} z^{q-1} + \dots + \beta_1 z^1 + \beta_0}, \tag{6}$$

wherein $\hat{O}(k)$ is the output provided by the estimated model $G_e(z)$, and $\mathcal{Z}$ represents the Z-transform operation. See that, $[\alpha_p, \alpha_{p-1}, \dots, \alpha_1, \alpha_0, \beta_{q-1}, \beta_{q-2}, \dots \beta_1, \beta_0]$ is the set of coefficients of $G(z)$ that the BiSI attack aims to discover, wherein p and q represent the order of the numerator and denominator, respectively. Therefore, to obtain the model of the actual device $G(z)$, the parameters of the estimated model $G_e(z)$ are modified and adapted until the output $\hat{O}(k)$ of $G_e(z)$ converges to $\mathcal{O}(k)$. To do so, the BSA iteratively adjusts the parameters of $G_e(z)$ by minimizing a fitness function f, until $G_e(z)$ meets $G(z)$. The coordinates $x_j = [\alpha_{p,j}, \alpha_{p-1,j}, \dots \alpha_{1,j}, \alpha_{0,j}, \beta_{q-1,j}, \beta_{q-2,j}, \dots \beta_{1,j}, \beta_{0,j}]$ of each individual j of the BSA are assigned as the coefficients of an estimated model $G_e(z)$. The fitness f_j of each individual j of the BSA is computed according to (7):

$$f_j = \frac{\sum\limits_{k=1}^{\tau} \left[\mathcal{O}(k) - \hat{O}_j(k)\right]^2}{\tau}. \tag{7}$$

Recall, from Fig. 3, that τ is the number of samples contained in each portion $o_n(k)$ of $o(k)$, and, therefore, is also the number of samples contained in $\mathcal{O}(k)$ and $\hat{O}_j(k)$. The signal $\hat{O}_j(k)$ is the output of $G_e(z)$ (6) when its coefficients are defined as x_j. From (7) it is possible to see that $\min f_j = 0$ if $\mathcal{O}(k) = \hat{O}_j(k)$. This result is achieved whenever $[\alpha_{p,j}, \alpha_{p-1,j}, \dots, \alpha_{1,j}, \alpha_{0,j}, \beta_{q-1,j}, \beta_{q-2,j}, \dots, \beta_{1,j}, \beta_{0,j}] = [\alpha_p, \alpha_{p-1}, \dots, \alpha_1, \alpha_0, \beta_{q-1}, \beta_{q-2}, \dots, \beta_1, \beta_0]$ or, in other words, when $G_e(z) = G(z)$.

Algorithm 2. BiSI attack with the noise processing strategy

---

begin
    Eavesdrop $i(k)$ and $o(k)$ during T samples;
    Noise Processing
        Select all i_n and the respective $o_n(k)$, $\forall i(k) > \Omega$;
        Compute $\mathcal{I}(k)$ and $\mathcal{O}(k)$ according to (3), (4) and (5);
    end
    Execute BSA, using $\mathcal{I}(k)$ and $\mathcal{O}(k)$ to find $G(z)$.
end

The Algorithm 2 briefly describes the complete BiSI attack with the proposed noise processing strategy. Albeit the BiSI attack herein proposed uses the same bio-inspired metaheuristic used in [9] (*i.e.*, the BSA, concisely described in Sect. 2 as in [9]), its is worth mentioning the differences from the present attack and the BiSI attack of [9]:

- In [9] the attacker injects an attack signal in the system to identify its transfer function. In that approach, the presence of noise affects the ability of the attack to learn the system model from the outputs caused by the attack signal. On the other hand, in the present work, the attacker does not injects an attack signal in the system. Conversely, the attacker passively collects the noisy signals and use them to estimate the system transfer function.
- The approach presented in [9] does not use the Noise Processing technique herein proposed.

4 Results

This section presents an evaluation on the performance of the BiSI attack with the noise processing strategy presented in Sect. 3. The model of the attacked device – *i.e.*, the device to be identified – is represented by (8). In practice, such second order transfer function can represent, for instance, a DC motor [4] or a lighting system [5] (among other systems). However, it is worth mentioning that, depending on the system characteristics, the coefficients of such plants can be different from the example defined by (8).

$$G(z) = \frac{\mathcal{Z}[o(k)]}{\mathcal{Z}[i(k)]} = \frac{2}{z - 0.9}. \tag{8}$$

The sample rate is 50 samples/s, and the noise measured in the input of $G(z)$ is a white gaussian noise $w(k) \sim N(\mu, \sigma)$, wherein N is a normal distribution with mean $\mu = 0$ and standard deviation $\sigma = 0.005$. This way, 95% of the amplitudes of $w(k)$ are within ± 0.01 (2σ).

The results of this section were obtained through simulations using MATLAB/SIMULINK. The evaluate the benefits – in terms of accuracy – provided by the noise processing technique described in Sect. 3, two BiSI attacks are implemented for comparison:

(I) a BiSI attack using the noise processing technique along with the BSA optimization process, such as described in Sect. 3;

(II) a BiSI attack using only the BSA optimization process (*i.e.*, without the noise processing stage). In this case, the eavesdropped signals $i(k)$ and $o(k)$ are directly used – without treatment – by the BSA to estimate the parameters of $G_e(z)$. To do so, Eqs. (6) and (7) – used to compute the fitness of BSA individuals – are rewritten as (9) and (10), and the BiSI attack is simply represented by Algorithm 3.

$$G_e(z) = \frac{\mathcal{Z}[\hat{o}(k)]}{\mathcal{Z}[i(k)]} = \frac{\alpha_p z^p + \alpha_{p-1} z^{p-1} + \dots + \alpha_1 z^1 + \alpha_0}{z^q + \beta_{q-1} z^{q-1} + \dots + \beta_1 z^1 + \beta_0}, \tag{9}$$

$$f_j = \frac{\sum\limits_{k=1}^{\tau} [o(k) - \hat{o}_j(k)]^2}{\tau}. \tag{10}$$

Algorithm 3. BiSI attack without the noise processing strategy

---

begin
    | Eavesdrop $i(k)$ and $o(k)$ during τ samples;
    | Execute BSA, using $i(k)$ and $o(k)$ to find $G(z)$.
end

As previously discussed, the BiSI attack aims to estimate the coefficients of the LTI transfer function of an NCS device. Therefore, in the present simulations, the parameters to be identified – according to (8) – are $\alpha_0 = 2$ and $\beta_0 = 0.9$. The BSA configurations in this paper are the same as those used in [9,10]: the lower and upper limits of each search space dimension are -10 and 10, respectively; the number of individuals in the BSA population is 100; $\eta = 1$; and the stopping criteria is 600 iterations. Moreover, $T = 0,5M\,samples$, $\tau = 100\,samples$ and $\Omega = 0.01$.

Each of the BiSI attack implementations – (I) and (II) – are evaluated through 31 different simulations. Each simulation uses a different white gaussian noise signal, randomly generated. Figure 4 shows the 31 values of α_0 and β_0 estimated by the two BiSI attack implementations (*i.e.*, with and without the noise processing stage). Additionally, Table 1 shows the statistics of the results presented in Fig. 4. From Fig. 4 and Table 1, it is possible to verify that the accuracy of the BiSI attack with the noise processing stage is better than the accuracy of the BiSI attack without the proposed technique. Figure 4(b) indicates that the two implementations have similar performance when estimating β_0. In both implementations, all estimated β_0 are close to the actual β_0 and, according to Table 1, the standard deviations are similarly low. On the other hand, Fig. 4(a) demonstrates that implementation (I) has better performance than implementation (II) when estimating α_0. With the noise processing stage, the estimated values of α_0 are closer to the actual α_0 – *i.e.*, less spread than without the noise processing stage. The statistics shown in Table 1 ratifies the better performance provided by the noise processing stage when the BiSI attack estimates α_0. In this case, the mean of the estimated values is closer to the actual α_0, with lower standard deviation.

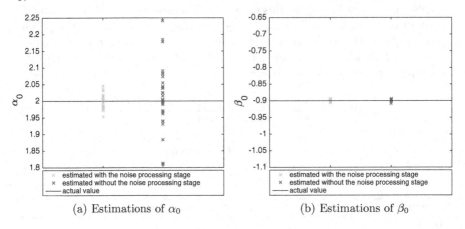

(a) Estimations of α_0 (b) Estimations of β_0

Fig. 4. Estimations of α_0 and β_0 with and without the noise processing stage.

Table 1. Statistics of the BiSI attacks

Coefficient	BiSI attack implementation	Mean	Standard deviation
α_0	(I)	1.9997	0.0189
	(II)	2.0119	0.0911
β_0	(I)	−0.8999	0.0034
	(II)	−0.8998	0.0024

Figure 5, obtained from one example of BiSI attack using implementation (I), compares the impulse response function $\mathcal{O}(k)$ of $G(z)$ – computed by the noise processing stage – with the impulse response function $\hat{\mathcal{O}}(k)$ of the estimated model $G_e(z)$. Note that, this figure demonstrates the product of the work done by the noise processing stage: a clear impulse response function, extracted from a white gaussian noise, that is better handled by the bio-inspired identification process performed by the BSA. It is possible to see how close $\hat{\mathcal{O}}(k)$ is from $\mathcal{O}(k)$, which demonstrates the high accuracy of the estimated model $G_e(z)$ when the BSA-based identification uses the signals provided by the proposed noise processing stage.

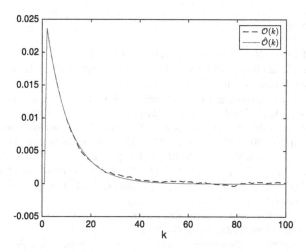

Fig. 5. Evaluation of the performance of the identification process – comparison between $\mathcal{O}(k)$ and $\hat{\mathcal{O}}(k)$.

5   Conclusion

In this work we propose a noise processing technique to improve the accuracy of bio-inspired system identification algorithms. The simulation results indicate that when the proposed technique is performed prior to the BSA-based system identification process, the accuracy of the estimated model increases. Therefore, the present technique represent a useful tool to make BiSI attacks effective in noisy NCSs. The proposed technique overcomes the constraint presented in other implementations of BiSI attacks, where the accuracy of estimated models used to be degraded by noise. The outcomes indicates that, with this approach, noise may not be a problem for a BiSI attack. Instead, noise can represent a meaningful and useful information for an attacker if he/she uses the approach described in this paper.

For future work, we plan to investigate techniques to mitigate BiSI attacks, by hindering the identification process in situations where an attacker has access to the data flowing in the NCS. Moreover, we plan to investigate the use of the proposed algorithm as a defense tool to identify possible model-based attacks in noisy NCSs. In this sense, we believe that this algorithm can be used to provide the NCS with information regarding the model of an eventual attack, in order to allow the autonomous reconfiguration of the control function to compensate the presence of the attack.

Acknowledgements. This work was partially sponsored by the EU-BR SecureCloud project (MCTI/ RNP 3rd Coordinated Call), by the Coordination for the Improvement of Higher Education Personnel (CAPES), grant 99999.008512/2014-0, and by FCT through project LaSIGE (UID/CEC/00408/2013).

References

1. Amin, S., Litrico, X., Sastry, S., Bayen, A.M.: Cyber security of water SCADA systems Part I: analysis and experimentation of stealthy deception attacks. IEEE Trans. Control Syst. Technol. **21**(5), 1963–1970 (2013)
2. Civicioglu, P.: Backtracking search optimization algorithm for numerical optimization problems. Appl. Math. Comput. **219**(15), 8121–8144 (2013)
3. Farooqui, A.A., Zaidi, S.S.H., Memon, A.Y., Qazi, S.: Cyber security backdrop: a scada testbed. In: 2014 IEEE Computing, Communications and IT Applications Conference (ComComAp), pp. 98–103. IEEE (2014)
4. Ferrari, P., Flammini, A., Rizzi, M., Sisinni, E.: Improving simulation of wireless networked control systems based on wirelessHART. Comput. Stan. Interfaces **35**(6), 605–615 (2013)
5. Ji, K., Wei, D.: Resilient control for wireless networked control systems. Int. J. Control Autom. Syst. **9**(2), 285–293 (2011)
6. Langner, R.: Stuxnet: dissecting a cyberwarfare weapon. IEEE Secur. Priv. **9**(3), 49–51 (2011)
7. Long, M., Wu, C.H., Hung, J.Y.: Denial of service attacks on network-based control systems: impact and mitigation. IEEE Trans. Ind. Inform. **1**(2), 85–96 (2005)
8. de Sá, A.O., da Costa Carmo, L.F., Machado, R.C.: A controller design for mitigation of passive system identification attacks in networked control systems. J. Internet Serv. Appl. **9**(1), 2 (2018)
9. de Sá, A.O., Carmo, L.F.D.C., Machado, R.C.: Bio-inspired active system identification: a cyber-physical intelligence attack in networked control systems. Mobile Netw. Appl., 1–14 (2017)
10. de Sá, A.O., da Costa Carmo, L.F.R., Machado, R.C.: Covert attacks in cyber-physical control systems. IEEE Trans. Ind. Inform. **13**(4), 1641–1651 (2017)
11. Smith, R.: A decoupled feedback structure for covertly appropriating networked control systems. In: Proceedings of the 18th IFAC World Congress 2011, vol. 18 (2011). IFAC-PapersOnLine
12. Smith, R.S.: Covert misappropriation of networked control systems: presenting a feedback structure. IEEE Control Syst. **35**(1), 82–92 (2015)
13. Snoeren, A.C., et al.: Single-packet IP traceback. IEEE/ACM Trans. Netw. (ToN) **10**(6), 721–734 (2002)
14. Teixeira, A., Shames, I., Sandberg, H., Johansson, K.H.: A secure control framework for resource-limited adversaries. Automatica **51**, 135–148 (2015)

Bio-inspired Approach to Thwart Against Insider Threats: An Access Control Policy Regulation Framework

Usman Rauf[1]([✉]), Mohamed Shehab[1], Nafees Qamar[2], and Sheema Sameen[3]

[1] Department of Software and Information Systems,
University of North Carolina at Charlotte, Charlotte, NC, USA
{urauf,mshehab}@uncc.edu, usman.cyberdna@gmail.com
[2] Governors State University, University Park, Chicago, IL, USA
mqamar@govst.edu
[3] IBM T. J. Watson, Yorktown Heights, NY, USA
sheema.sameen@ibm.com

Abstract. With the ever increasing number of insider attacks (data breaches) and security incidents it is evident that the traditional manual and standalone access control models for cyber-security are unable to defend complex and large organizations. The new access control models must focus on auto-resiliency, integration and fast response-time to timely react against insider attacks. To meet these objectives, even after decades of development of cyber-security systems, there still exist inherent limitations (i.e., understanding of behavioral anomalies) in current cyber-security architecture that allow adversaries to not only plan and launch attacks effectively but also learn and evade detection easily. In this research we propose a bio-inspired integrated access control policy regulation framework which not only allows us to understand anomalous behavior of an insider but also provides theoretical background to link behavioral anomalies to the access control regulation. To demonstrate the effectiveness of our proposed framework we use real-life threat dataset for the evaluation purposes.

1 Introduction

The ultimate goal of cyber-security and forensics community is to deal with wide range of cyber threats in (almost) real-time conditions. Whereas in today's cyber infrastructures, current Monitoring and Analysis (M&A) technologies (solutions) only address the facet of plethora of problems. The focus of current state of the art is towards developing Security Event Management (SEM) or Security Information and Event Management (SIEM) systems, which can efficiently collect event related information (in the form of logs) from operating systems or network devices (e.g., firewalls). This plethora of information is then used for analytics by a centralized unit for detection of malicious activities via signature-based testing

A. Compagnoni et al. (Eds.): BICT 2019, LNICST 289, pp. 39–57, 2019.
https://doi.org/10.1007/978-3-030-24202-2_4

or correlations between events. Finally, the alarm is generated to update the security personal, who is not only responsible for checking the legitimacy/correctness of alarm, but also have to implement security policies in manual/semi-automatic ways, to cope with real time situations. These technologies (SEM/SIEM) perform well, when it comes to aggregation of information, but due to the inherent limitations, do not provide any hint about ongoing insider attacks and policy synthesis mechanism against a legitimate user turned into a malicious user or insider attacker. These inherent limitations include lack of interaction between detection units and policy synthesis procedures, which results into an inability to synthesize security policies dynamically, via risk assessment or behavioral analysis. Next generation technologies are expected to eradicate these limitations by efficiently integrating modern technologies into a standalone monitoring and detection unit that can be deployed at various nodes in a network. Whereas very little or closer to none efforts are spent in developing technologies that can have the capability to not only detect user/entity abnormal behavior, but also to autonomously react against cyber threats in near real-time conditions using actionable information (threat intelligence).

Biological systems, on the other hand, have intrinsic appealing characteristics as a result of billions of years of evolution, such as adaptivity to varying environmental conditions, inherent resiliency to failures and damages, successful and collaborative operation on the basis of a limited set of rules. Inspired by the nature of cellular regulation mechanism, which helps to maintain an optimal concentration of proteins in a cell via signal transduction mechanism and mitigating the perturbations (insider threats) due to the high/low of productions of certain proteins, we present cellular regulation inspired (integrated) systemic approach through which the security policies can be dynamically altered against an originating threat (via detection and threat analytics). The threats under consideration are behavior anomalies, which make it difficult for the existing technologies (SEM/SIEM) to defend against an insider who is also a legitimate user, as there is no available benchmark or standard, other than guidelines, to differentiate a normal user from an abnormal/anomalous user.

Therefore, our first major contribution is to set forth the criteria to detect behavioral anomalies in near real-time situations. Our second main contribution is to formally model policy regulation problem as state transition system such that the formal tool can be leveraged for policy regulation/synthesis proposes. Finally, our third main contribution involves bio-inspired framework for the integration of threat analytic (for behavioral anomaly) and policy regulation attacks.

2   Related Work

Although the nature of insider threats differs from those of external attacks, the detection techniques can still be characterized in two categories: signature-based, or anomaly-detection methods [26]. Insider threat detection faces unique challenges as compared to challenges faced by detection of external attacks.

This uniqueness is due to many nontechnical factors which contribute towards the change in the behavior of employees, which have legitimate access to the resources with organizational knowledge. Hence, understanding of the change in behavior is of utmost importance to deal with insider threats, given the statistics of insider incidents [9,16].

2.1   Signature-Based Insider Threat Detection Systems

In case of insider attacks the signatures are defined in terms of predefined policies, which trigger alarm once violated, e.g., unauthorized access to a machine or file, for which access policy is predefined. For instance, Agrafiotis et al., develop a tripwire approach to detect actions that are indicators of insider threat based on designed policies on alarming behaviors, and attack-patterns [11]. The authors do not provide any experimental results, or the details about how their proposed approach will reinforce the policy regulation. IBM uses a similar approach as part of the IBM QRadar SIEM solution through the implementation of offences. Offences are designed to detect threats in general and may be used for detecting steps of known insider attacks [1]. Bishop et al., proposed a different approach to detect insider attacks by developing a solution based on the attack trees [13]. The authors build attack graphs to illustrate possible scenarios through which a target can be compromised by an insider. Finally, they determine Minimum Cut Set (MCS) to find out possible countermeasures for an ongoing attack. The proposed approach is highly dependent on the successful design of a process model (attack tree) that identifies the vulnerabilities of the process and possible attack targets. It is also limited to detecting attacks on the proposed targets as only known vulnerabilities can be modeled using attack trees, and the attack tree based approaches do not take into account any change in insider's intention/behavior. The authors also do not discuss how the detection will reinforce policy regulation mechanism.

2.2   Anomaly-Based Insider Threat Detection Systems

These systems are designed to detect unknown types of attacks and behaviors, and trigger an alarm if encounter any deviation. Some detection-solutions consider non-technical indicators of insider threat. Non-technical indicators, such as the psychological state of the insider are of crucial value to insider-threat detection [10,31]. Therefore, work has been done to incorporate their analysis in insider-threat detection systems. For example, Brdiczka et al., used graph anomaly detection and additional techniques to learn the normal behavior of nodes. They also use psychological profiling to take into consideration an insider's intention, which is a non-technical indicator, with the aim of reducing false positives generated by monitoring technical indicators [15]. The authors use gaming community data, World of Warcrafts (WoW) and social network-based activities, to analyze the behavior of a player in a group. The main limitation of the approach is that the testing data and attributes used for analysis have almost no relevance when it comes to insider threats in an organization, since the settings in which individual react in a character gameplay forum are much different than the malicious activities pattern of a legitimate insider. Finally, they do not present any methodology to map measured threat impact to the policy regulation of an organization.

Chen et al., proposed a belief-based threat detection system which also considers the intention of an insider as an indicator [19]. Their solution is designed to estimate the probability of success of an attack by conducting behavioral analysis using probabilistic model checking. Prediction is done after a potential insider has been identified through intentional analysis using Bayesian networks. The major flaw of the approach is the probability distribution calculation, and modeling of individual threat scenarios as Markove Decision Process (MDPs). The approach becomes highly unrealistic since it requires the modeling of each individual and a certain threat to be modelled and analyzed separately. In an organization of thousands of employees, the analysis becomes highly impractical. Second issue is the assignment of probabilities to the transitions in MPD model, which are unrealistically obtained and follow random Bernoulli distribution. The user based technical attributes are also mentioned which can help understanding user's behavior.

Brdiczka et al., and Chen et al., apply automated analysis of non-technical indicators of insider threat requiring the collection of sensitive data, such as the contents of email communications to be used for sentiment analysis [15,19]. Although their proposed methodology promises to deliver high accuracy but the maximum accuracy of detection system is only 82%. Given the low accuracy, technical challenges related to the deployment and lack of understanding of users behavioral impact on policy regulation, makes this approach impractical and an unviable choice.

Some anomaly-detection methods are developed to detect a certain type of insider threat. For example, Zhang et al. [32], propose a solution to analyze document-access behavior to classify users based on the contents of accessed documents. Each user is identified by the type of documents they usually access. Anomaly detection checks for deviations from historical and current behaviors of the user, and the behavior of the community using the Naive Bayes algorithm and correlation matrices. This approach is limited to monitoring a single indicator (type of accessed files), referring to a specific type of insider threat, for instance information leakage. Whereas according to the CERT guide and SANS survey to insider threats combining multiple indicator can provide better detection efficiency [8,16].

Other detection methods aim at detecting threats to a specific resource in an organization. For example, Senator et al., develop a solution to detect threats to database-access behavior. Their solution is an example approach that is limited to protecting a specific resource (corporate database). Although the authors consider multiple indicators for anomaly-detection algorithms to tackle with the low signal-to-noise ratio challenge in insider threat, they do not provide any information about the impact of detection unit's output on policy regulation, and their approach assumes that the actions will be taken by an analyst [37].

Finally, some detection methods are designed to learn a normal behavior of employees from their online activities. For instance, a more relevant work includes the approach presented by Legg et al., [28]. They developed an automated detection system that uses Principle Component Analysis (PCA) to detect anomalies. They compute hourly feature vectors on the activities of employees and build a 24-hour matrix of activities. Then, PCA is applied to project the multivariate vectors into a 2D space based on the maximum variance exhibited by features. Anomaly detection then measures the distance of points in the projected space from the origin. The variance based anomaly-detection method is difficult to interpret as the classification requires predefined threshold, this limits a security analyst's capability of gaining insight on the decision-making process of the method while investigating the generated alarms. Another issue is that the approach clusters the users together, which makes it harder for policy integration, as policy is defined for a malicious activity not a whole group.

Rashid et al., make use of Hidden Markov Models to learn the normal behavior of employees and analyze deviations from the learned behavior to detect insider threat, and consider normality as a sequence of events [33]. The authors highlight that their model offers the advantages of learning parameters from the dataset that describe an employee's behavior. Their model is also advantageous in learning from data that is sequential in nature. However, the computational cost of training the models increases as the number of states captured increases, while the effectiveness of the method in detecting insider threat is highly impacted by the number of states. Although the proposed approach is able to capture the anomalous behavior of an insider, it does not provide how efficiently it can report to an analyst and how the detection results can be mapped to policy regulation. Moreover, Song et al., use Gaussian Mixture Models for modelling the behavior of users for insider threat and masquerade detection. They compare their results with several other machine learning methods based analysis and find it superior in achieving higher accuracy values. However, their model is applied on system-level events, such as process creation, intended for a biometric identification instead of identifying user behavioral anomalies [36].

2.3   Policy Regulation in RBAC

Although RBAC is the most widely used access control architecture which has several benefits, it cannot automatically revoke users' access if they are on the verge of behaving maliciously or not behaving properly. For this reason, several approaches have incorporated the notion of trust in RBAC [17,23,25]. However, existing approaches neither present a comprehensive analysis of the way in which trust thresholds should be assigned, nor specify how to enforce such policies or reduce the risk exposure automatically. In [17], roles are associated with trust intervals, and trust intervals are assigned to users. Users are assigned to roles according to their trust levels. This model does not capture the intuitive nature of RBAC systems in which users are assigned to roles according to their organization's functions, not trust levels. In [25], users are assigned to roles based on trustworthiness and context information. A similar approach was proposed in [18], where role thresholds are a function of the risk of the operations. If the trust of the user offsets the risk of the action, the access is granted. However, none of the existing works provide a clear understanding about trust computation and do not provide any method to reduce the risk that an organization faces at runtime by selecting roles with minimum risk exposure.

In the research proposed by Ma et al. [29], each role is assigned a minimum level of confidence and each user is assigned a clearance level. Based on these values, the risk associated with a user activating a role is calculated. Objects and actions are assigned a value according to their importance and criticality. However, this work does not mitigate insider threats as the trustworthiness of users is defined as a static parameter that does not depend on users' behavior. In addition, the authors do not consider role hierarchy in their work and do not present experimental results [29].

In [12, 18, 30], the main focus is also to reduce the risk exposure. In [30], a risk based analysis is proposed to ensure that system administrators assign permissions to the roles considering the risk inherent to those permissions. Each permission is assigned a risk value, and the role hierarchy is organized based on these risk values. This may not be appropriate, as it is more intuitive to organize the role hierarchy according to the employee's structure. We argue that maintaining a role hierarchy that matches the organization's hierarchy is more intuitive for security administrators. Additionally, this work does not reduce the risk exposure of the organization during the role activation process.

In [12], a model that modifies the policy to minimize the risk exposure as systems evolve is proposed. The risk is considered as a parameter which varies in an interval over $[t, t') \in \mathbb{R}$. User is assigned obligations on the basis of assessed risk. The approach do not provide any understanding about risk calculation, behavioral anomaly, and obligation fulfillment check. Since, the proposed system can assume any arbitrary state given the values of risk and obligation attributes, it makes management (adding, removing or updating policies) much more harder for an administrator, as there is no way to comprehend underlying state of the system, making it cumbersome to modify it and prone to errors.

Chen et al., propose a model in which the risk associated with a role is calculated using the trustworthiness of the user, the degree of competence that a user has to activate a role, and the degree of appropriateness of the permission-role assignments. Each permission is assigned a mitigation strategy, which is a list of risk thresholds and an associated obligation pair [18]. When the user wants to obtain a set of permissions, the role with minimum risk is selected. Then, the system consults the mitigation strategy to see which action is more appropriate: to deny the access or to allow the access imposing an obligation. The authors, do not account for the context as an important component to define the risk threshold that should be enforced.

Salim et al., propose to assign costs of access to permissions depending on the risk of their operations, and to assign to each user a budget [35]. Users are assigned roles, but being assigned or not does not necessarily determine whether or not a user should be allowed to activate a role. If the user accesses permissions that he/she can obtain through an authorized role, the cost is reduced. In case the user is not authorized to a role, the cost of activating the role is taxed. Nonetheless, if the user has enough budget to make the operation, he/she can access the permissions. The authors claim that this mechanism incentivizes users to spend their budget cautiously, activating low cost (low risk) roles. However, this scheme exacerbates the risk of insider threats. Users can use their budget to perform unauthorized accesses without being detected; e.g., if a disgruntled employee wants to quit the organization, he/she would not mind expending all his budget performing a malicious action.

Many commercial products also incorporate risk in their solutions; e.g., SAP [3], Oracle [4], IBM [6] and Beta Systems [5]. These products mitigate risk by closely monitoring and auditing the usage of risky permissions. The risk values, however, are not used to make access control decisions, missing the opportunity to incorporate the overall known behavior of the users to prevent insider threats. The threat of inference of unauthorized information is particularly relevant in the insider threat context. This threat occurs when through what seems to be innocuous information, a user is capable of inferring information that should not be accessible. In existing approaches to deal with inference threat [14], when the user is about to infer some unauthorized information, the system prevents it by either denying access or providing scrambled data. This is not adequate for all types of organizations. We believe that real organizations may need to provide access to multiple pieces of information to a single employee even if they result in undesirable inference. Existing RBAC extensions do not consider the risk of inferred information. New ways to mitigate the inference risk in RBAC-based systems are needed and this forms our motivation as in this article we intend to develop a integrated detection and response systems which can make use of threat intelligence to autonomously regulate access control policies. Towards this direction, in the next section we discuss our (bio) inspiration system in details.

3   Cellular Regulation via Signal Transduction

Every functionality in the human body and evolution of the morphological features is highly influenced or controlled at (intracellular) molecular level [20,34,38]. Genes and proteins are the main ingredient of this controlling mechanism, which play together in a programmed manner to perform multiple tasks in an organism. Genes are the informative subunits of DNA and they decode instructions in the form of proteins. When a gene is switched on, information flows from genetic to proteomic level as complex processes of transcription and translation. Some proteins have the function of regulating the expression of genes by turning them on or off. This complex interactions of genes and proteins to regulate the cell against any external or internal threat/perturbation after receiving signals from cellular receptors is referred as Cellular Regulation via Signal Transduction [34]. Regardless of where the control comes from, whether its hard coded in DNA or nucleolus of the cell, in cellular regulation, the key principle is the regulation of different parts of DNA (resulting in controlled synthesis of different proteins) against any uprising threat at cellular level.

As a first step towards creating bio-inspired resilient architecture, we intend to understand how this feedback notion works and can help us in accomplishing our objectives. In the next section, we present a real life biological phenomenon for the better understanding of readers.

3.1 Blood Pressure Regulation System

Renin angiotensin system (RAS) plays a crucial role in physiological functioning of human body by regulating blood pressure. This hormone control system is triggered to avoid the drop of blood pressure towards some critical life threatening level in different stress conditions e.g., dehydration and hemorrhage.

In human body the decrease in blood pressure is primarily sensed by specialized cells in kidneys which increase the production of Renin enzyme as a consequence. Renin catalyzes a protein called Angiotensinogen, which is produced by liver, into another protein angiotensin I. Angiotensin I is further converted in to Angiotensin II by angiotensin converting enzyme (ACE). Angiotensin II is the main product of RAS system which increases blood pressure by a triple action plan: (1) it constricts blood vessels in kidneys by contraction of smooth muscle, cells (2) it enhance the production of aldosterone hormone which helps in Na+ retention in kidneys, and (3) triggers the production of vasopressin hormone in the brain. All these three actions performed by angiotensin II are essential for blood regulation in body.

The angiotensin II protein performs all three tasks by first binding to the receptors of target cells. The binding of angiotensin II with receptors trigger cascade of biochemical reactions which result in aforementioned tasks responsible for elevation of blood pressure. The reaction stops eventually when all the receptors are bound by the protein. The rennin secretion also stops due to increase in blood pressure up to normal levels and hence it also blocks the conversion of angiotensinogen to angiotensin II.

The cause for hypertension or high blood pressure is hidden somewhere in the RAS system. As the angiotension II is the key functional element of this system so most of the therapies are designed to control this protein by blocking its activity which is done by blocking of ACE enzyme, responsible for the conversion of angiotensin II from angiotensin I. The drugs targeting ACE, also known as ACE inhibitors, has shown promising results in therapy of high blood pressure disease. For detailed information and discussion about RAS, we divert our readers to the article presented by Dressler [24]. In the next section we summarize the working procedure of cellular regulation in the form of a framework, and propose similar framework for regulation of insider threats, inspired by the phenomenon of cellular regulation.

4 Cellular Regulation Inspired Mapping and Proposed Framework

To proceed further with the idea of integrating auto-resiliency characteristics of biological systems in current cyber architecture, there must exist some analogy and mapping of actions among fundamental entities of both domains. Table 1 illustrates mapping of cellular regulation principles towards cyber-security concepts. This also forms the basis of our proposed framework (c.f. Fig. 1).

Table 1. Analogy between cellular regulation and cyber-security

Biological characteristics	Cyber characteristics
Fundamentals	
DNA: is a conjunction of regions, which are individually represented as genes. These regions, under any perturbation, are regulated (turned on/off) via complex processes to neutralize any threat that cells may face	Cyber Policies, on the other hand, are conjunction of rules, which can be added or removed (activated or turned off), to deal with any potential threat to an organization
Sensing Mechanism	
At cellular level, a cell's receptors are responsible for receiving signals about ongoing activities and traverse them within a cell. Nature has fine tuned sensing mechanism through which elements within a cell are made aware about the changing environment around them	Log collection mechanisms in cyber domain, on the other hand, can be tuned to collect information about users and their activities at system/network level in real time. These mechanisms (if implemented in real time) can perfectly mimic and correspond to sensing mechanism in cellular regulation
Actuation/Regulation	
At cellular level, concentration of a protein is changed via turning on/off its regulating genes (regions of DNA), against any perturbation	If we consider policy as a disjunctive conjunction of all possible rules, then dynamic selection of a suitable subset of rules can be referred as turning on/off regions of a policy, to make it more appropriate according to the current scenario and perfectly corresponds to the regulation at cellular level

In the next step, we propose cellular regulation inspired framework to deal with insider threats. As we established earlier, although the cellular regulation process encompasses and neutralizes both internal and external threats, but our focus in this article is toward developing an architecture which can deal with insider threats. The motivation of choice is due to the limitation in availability of the testing datasets. Although with minor modification the proposed framework can also be extended to deal with external threats, but we limit our focus to

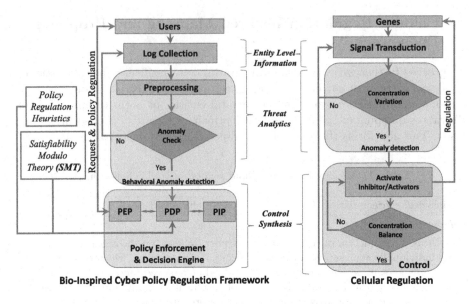

Fig. 1. Mapping of cellular regulation mechanism to the proposed cyber policy regulation framework

the internal/insider threats only, as we can observe the variations in insider's behavior via activity log collection. Figure 1 presents a detailed description of our proposed framework along with side by side comparison to the working principles of Cellular regulation process. We discuss each component in details in the forthcoming discussion.

4.1 Sensing Module

As we established earlier that in cellular regulation, sensing is performed by cell's receptors, which monitor continuously the concentration levels of proteins, and disperse that information internally (within a cell), so that various genes can be made aware of the perturbations in their environment. In our case we propose to collect event logs of user's activities, and pass them to a central entity (Threat Analytics Module) for threat analysis. Rather than dispersing information to every entity in the system (as in case of cellular regulation), we propose to collect logs in centralized manner, as our control (policy regulation) originates from central authority (Policy Enforcement & Decision Engine).

Although the log collection process for threat analytic is not a novel concept, but it exists in individuality and in this research we propose to integrate it with access control mechanism by following the principles of cellular regulation. Most of the recent recommendations propose active log collection mechanisms but none of them provide any guidelines about how to make activity logs information meaningful and usable for threat analytic using machine learning methods.

4.2   Threat Analytic Module

Feature Engineering: As a first step towards threat analytics, one of the main contributions of this article is to setup a criteria for pre-processing/feature-engineering of the data (logs) collected from the sensing module. Event related data is stored in the logs with time stamps, identifying activity performed by an individual at certain time. Activities under consideration can vary depending on organizational needs, but most generic activities include: login/logoff details, web surfing, use of plugin devices, and duration of specific activities.

The first challenge is to convert these event based time-labeled data in to a meaning full dataset. Which can be utilized by machine learning methods. As a first step we convert time-stamps into a cyclic temporal variable which varies between 0–24 h. This makes it easier to assign a numeric value to any event and allows machine learning classifiers to learn and tie an activity with a number rather than an uninterpretable string (date and time). The second main issue we address during feature engineering phase is the separation of each individuals data, so that we can train classifiers over an individual's behavior. The sample of the updated dataset with temporal encoded information can be found in [2].

Possibility of being able to learn an individual's behavior (from activity logs) allows us to measure the variation in it as well. Once the deviation of an individual from its own (and others) is predictable, we can easily compare predictions with ongoing activities. For instance, if an employee logs in during a certain time window over a course of time, then using machine learning methods, classifiers can be trained to learn about the login time window slot and predict in which time slot an individual mostly/normally login and starts working. We use OneHotEncoding method to encode the employees data and separate each individual's data from the log files. Finally we combine all employees data to construct a dataset which is finally usable by machine learning methods [2].

Anomaly Check: Having an abnormal work routine do not refer to an anomaly. Therefore, we do not consider abnormality as an anomaly. An individual may have different work patterns, and may be abnormal but as far as the high risk activities and checks are not triggered, we consider a user to be just abnormal and not anomalous. The best way to find out anomaly in this situation is to compare an individual's profile with its own working routine and observe significant and abrupt variation. For instance, an employee has not used SSH connection to a secure data repository at midnight, in the recent month, but suddenly has established SSH connection and is trying to upload data to a remote server around mid-night. We consider this type of variation in behavior as anomalies. In our proposed framework we use machine learning to train models against a user's profile (long term or short term behavior), and then predict there activities every time they try to access resources. If our predictions do not match the ongoing activities, threat analytic module considers it as user behavior anomaly and reports it to the Policy Enforcement & Decision Engine. We discuss the details of accuracy of our behavioral anomaly detection unit in the forthcoming Sect. 5.

4.3   Policy Regulation Module (PRM)

The third and most important component of our proposed *Bio-inspired Policy Regulation Framework* is *Policy Enforcement & Decision Engine*. It consists of three sub-modules: Policy Enforcement Point (PEP), Policy Decision Point (PDP), and Policy Information Point (PIP). PIP contains information about organizational policies (e.g. given user attributes what level of access can be granted). PEP receives user's attribute and passes it onto PDP. PDP utilizes PIP knowledge-base as a reference point and makes decision about granting/revoking access against a certain user/insider. We propose to integrate PDP with behavioral anomaly detection unit, so that they can operate autonomously, and regulate access control without human intervention.

In order to work autonomously, PDP requires threat information regarding an insider/user which generates access request and on this basis it decides whether or not the access should be regulated. There are multiple ways to implement this integration. (1) PDP can inquire about a user's threat level from anomaly detection module, or (2) anomoly detection unit can update threat levels and push these details into the PDP module. Since, detection and regulation modules are working independently (other than the dependency of PDP for threat levels), we propose the later option, to avoid any excess overhead. Although, PDP has the tendency to make decision about access control, but the current architecture of PDP lack the notion of understanding about threat and synthesizing access control against it. Towards this direction we formalize access control problem as constraint satisfaction problem, and use Satisfiability Modulo Theory (SMT) solver to solve it [21]. Use SMT allows us to enhance the capability of PDP to understand notion of threat and solve the problem of access regulation using constraint satisfaction concepts.

Before allowing access to a user's request, PRM checks if the threat level has altered or not. If threat level changes the decision engine either revokes or limits the access of a user/insider, according to the organizational requirements. Decision engine can then be integrated with the policy enforcement module to enforce the policy.

We formally define Policy Regulation Module as a transition system. A *Policy Regulation Transition System (PRTS)* can be defined as a *8-tuple*:
$\mathcal{M} = (S, s_0, u_i^r, A_j^c, P_i, T_{ij}^k, \hookrightarrow, \delta)$:

- S is a finite set of states of a policy (possible configurations) with cardinality in $\mathbb{N}$;
- $s_0 \in S$ is the initial/current state/configuration of policy;
- u_i^r is the rank of user i with values in $\mathbb{N}$;
- A_j^c is the confidentiality level of asset j with values in $\mathbb{N}$;
- T_{ij}^k is the threat level/impact of a given request by user i to access asset j, which can be calculated as:

$$T_{ij}^k = L_i \times Imp_j$$

whereas, L_i represents the likelihood of behavioral anomaly for a give user (i), and Imp_j is the impact if the asset (j) is being compromised.

- dec_k^{ij} is decision variable in $\mathbb{B}$, which helps SMT finding a new state for transition;
- $C(\mathcal{T}_{ij}^k)$ is a set of constraints over the threat vector/values;
- $\hookrightarrow$ is a finite set of transitions such that: $\hookrightarrow\subseteq (S \times \mathbb{N} \times \mathbb{N} \times \mathbb{B})^2 \times C(\mathcal{T}_{ij}^k)$;
- δ is a finite set of transition rules which maps $C(\mathcal{T}_{ij}^k)$ to set of transition $\hookrightarrow$;

We define security policy as disjunctive conjunction of rules, and rules (r_{ij}) contain information about user (u_i), and requested asset (A_j).

$$\mathcal{P} : \bigvee_k (\bigwedge_{ij} (r_{ij})) \ whereas, i, j, k \in \mathbb{N}$$

The above mentioned expression shows the assumption that there exist all possible combination of the rules (we call configurations of a policy) in the knowledge-base, and given this assumption PRTS can switch configuration under the effect of information provided by Threat Analytics Module as per the following semantics.

$$(s, u_i^r, A_j^c, \mathcal{T}_{ij}^k, dec_k^{ij}) \xrightarrow{\mathcal{T}_{ij}^k >= \theta; dec_k^{ij} = 0} (s', u_i^r, A_j^c, \mathcal{T}_{ij}^k, dec_k^{ij'})$$

$$(s, u_i^r, A_j^c, \mathcal{T}_{ij}^k, dec_k^{ij}) \xrightarrow{\mathcal{T}_{ij}^k < \theta; dec_k^{ij} = 1} (s'', u_i^r, A_j^c, \mathcal{T}_{ij}^k, dec_k^{ij''})$$

The above mentioned formalism defines the semantics of our proposed PRTS. Constraints over $(\mathcal{T})$ work as guards over transitions $\hookrightarrow \subseteq (S \times \mathbb{N} \times \mathbb{N} \times \mathbb{N} \times \mathbb{B})^2$. The transition from one configuration to another configuration is only fired once the guards evaluate to true and invariants are violated (threat level of a user increases). Whereas, s, s' and s'' are distinct states such that s and $s' \in S$ and $s \cap s' : \emptyset$. If the threat is below a transition triggering threshold s and s'' may or may not be the same. Once the threat impact is evaluated by detection unit, the new configuration is selected by SMT solver by solving the following constraint:

$$\exists_{i,j,k} \left(\bigvee_{k:1}^n ((\bigwedge_{i,j:1}^m (r_{ij})) \bigwedge (dec_k^{ij})) \right) \ whereas, i, j, k \in \mathbb{N} \tag{1}$$

The above expression only finds the configuration of the policy in which dec_k^{ij} is true or 1 (as it is a binary decision variable), which means allowing only the configurations in which the threat is under acceptable threshold and values of i,j,k does not necessarily have to be equal. For instance given a current state s if the value of $\mathcal{T}_{ij}^k$ becomes higher than the acceptable threshold (θ), as per the analysis provided by detection module, transition will occur as per the above mentioned transition semantics, and new state will be selected by solving the above mentioned constraint 1. In the following section we discuss the evaluation and effectiveness of our approach.

5   Evaluation of Bio Inspired Policy Regulation Framework

5.1   Effectiveness of Behavioral Anomaly Detection Unit

To measure the accuracy of our behavioral anomaly detection unit, we use threat test data set released by CERT in 2016 [7,27]. The dataset contains the log activities for one thousand employees and contains five different types of insider threat scenarios, for detailed description of the scenarios we refer our reader to the dataset details (*file: scenario.txt*) in [7] . In the context of this paper, we only focus on the first scenario, and use information of the dataset which is relevant to this scenario. Although our approach can be used to deal with other scenarios, we only use scenario one for testing and evaluation purposes.

Scenario: *User who did not previously use removable drives or has variable routine begins logging in different hours, using a removable drive, and uploading data to a listed malicious website.* After prepossessing of the provided user activity logs

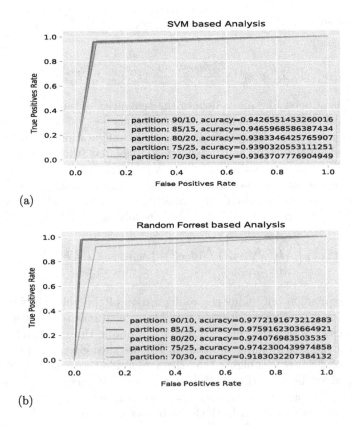

(a)

(b)

Fig. 2. Evaluation of train-test sample size partitioning to find optimized partitioning size

[7], we construct our own threat test dataset which can be accessed and used for machine learning purposes [2]. As we mentioned earlier we use OneHotEncoding so that the information regarding employees which are to be considered for analysis should be in separate columns as depicted in our processed dataset. Each attribute, for example login time of a day, and activity performed (use of external hard drive, or visit to malicious website) are represented in separate columns. Values of time attribute are mapped between 0–24 h range, whereas the values of activities are binary, representing "1" if an activity is triggered by a certain employee at a certain time, and "0" otherwise.

We use state of the art Random Forrest and SVM methods for predictions about insider's behavior. Following RoC curves and accuracy labels show that we were able to achieve almost 98% accuracy. Which means we were able to predict the behavior of an insider with high accuracy. In data science and machine learning the accuracy of analysis is highly dependent on two factor, (1) partitioning size of dataset while training the prediction classifier and (2) temporal variation in data sample size (e.g. variation in number of weeks). We vary both of these factors to observe the impact and find the optimized values for number of weeks to be considered for effective predictions and optimized size of partitions for training-and-test purposes.

Effect of Variations in Train-Test Partitioning Size. RoC curves in Fig. 2 show how the effectiveness of our behavioral detection unit varies with the variation in partitioning. We deduce from our analysis that Random forest based predictions were more accurate than SVM based predictions, and we were able to achieve ≈98% accuracy. We also observe ideal cutoff point for the train-test split to be 75%/25% by variation of train-test split ratios, since we do not observe any significant improvements by increasing the training set size.

Effect of Temporal Variations. Figure 3 shows how the effectiveness of behavioral detection unit varies with Temporal data size variation. For instance, given a general perception that having large data size or training a model over data dispersed over larger period of time helps to increase the accuracy of the results. We find it contradicting in case of behavioral anomaly problem. Our results show that the effectiveness of behavioral anomaly detection decreases rapidly if we train our model over the logs of larger period of time. Which means if we consider two week's logs (of an employee) for training and prediction, the accuracy will be higher than the scenarios in which we consider the logs of five weeks. We believe the degradation in the effectiveness is due to the over approximation of the models leading to higher false positive values. This type on analysis, helps us setting up a benchmark over the estimation and prediction parameters. Hence, we deduce that 2-to-3 weeks logs are ideal for training and prediction purposes.

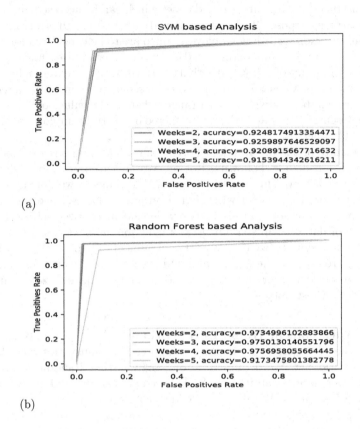

(a)

(b)

Fig. 3. Temporal evaluation to find optimal number of weeks for prediction analysis

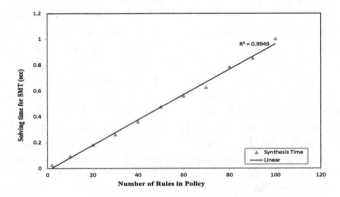

Fig. 4. SMT time to calculate satisfiable configuration of security policy

5.2   Efficiency of Policy Regulation Module

We also conduct a preliminary evaluation of the proposed *Policy Regulation Module* to assess its time complexity against number of rules in a policy, given the information from *Threat Analytic Module*. To compute satisfiable configuration of access control policy, we use Z3 SMT solver [22]. All the experiments are conducted on Core i5 machine with 2.4 GHz processor and 16Gb of memory. The result presented in the Fig. 4 shows how the time require to compute satisfiable configuration varies with the increase in the number of rules (in a policy). The result shows that the complexity is near to linear. In future, we aim to conduct extensive analysis about deployment of *Policy Regulation Engine* on larger scale.

6   Conclusion and Future Work

In this paper, we present a novel *Cellular Regulation inspired Access Control Policy Regulation Framework*, which not only observes the variation in the environment (behavioral anomalies of insiders), but also triggers necessary responses (policy regulation) to secure the system. Our first main contribution is to setup a feature (activity) based temporal criteria for understanding behavioral anomalies. The proposed integrated framework then utilizes the state of the art machine learning methods for the detection of behavioral anomalies. Our third main contribution is to model policy regulation problem as state transition system, which can utilize the results from behavioral anomaly module to refine the access control policy. We evaluate the efficiency and effectiveness of our proposed framework on real-life threat dataset provided by CERT. Our evaluation illustrates that we were able to achieve the accuracy of 98% (92% in worst case scenario) in case of behavioral anomaly detection. The evaluation also shows that we were able to synthesize the regulated policy in linear time for small set of rules. To the best of our knowledge, this is the first effort towards dealing with insider threats by unifying detection and deterrence systems. In future we aim to test our proposed system on medium to large scale examples for rigorous evaluation and incorporate more parameters for behavioral prediction to achieve high accuracy.

References

1. IBM QRadar, SIEM
2. www.dropbox.com/s/rerwekvuji12icm/logon_hotencoded_cleaned_data.csv?dl=0
3. Access risk management. Technical report (2012)
4. Application access controls Governor. Technical report (2012)
5. Identity and access Governance. Technical report (2012)
6. Resource access control facility (RACF). Technical report (2012)
7. CERT threat test dataset. CERT (2016)
8. Defending against the wrong enemy. Technical report, SANS Insider Threat Survey (2017)
9. Insider threat report. Technical report, CA Technologies (2018)

10. McCormac, A., Parsons, K., Butavicius, M.: Preventing and profiling malicious insider attacks. Technical report, Defense Science and Technology Organization, April 2012
11. Agrafiotis, I., Erola, A., Goldsmith, M., Creese, S.: A tripwire grammar for insider threat detection. In: Proceedings of the 8th ACM CCS International Workshop on Managing Insider Security Threats, MIST 2016, pp. 105–108. ACM (2016)
12. Aziz, B., Foley, S.N., Herbert, J., Swart, G.: Reconfiguring role based access control policies using risk semantics. J. High Speed Netw. **15**(3), 261–273 (2006)
13. Bishop, M., et al.: Insider threat identification by process analysis. In: 2014 IEEE Security and Privacy Workshops, pp. 251–264, May 2014
14. Biskup, J.: History-dependent inference control of queries by dynamic policy adaption. In: Li, Y. (ed.) DBSec 2011. LNCS, vol. 6818, pp. 106–121. Springer, Heidelberg (2011). https://doi.org/10.1007/978-3-642-22348-8_10
15. Brdiczka, O., et al.: Proactive insider threat detection through graph learning and psychological context. In: 2012 IEEE Symposium on Security and Privacy Workshops (SPW), pp. 142–149 (2012)
16. Cappelli, D.M., Moore, A.P., Trzeciak, R.F.: The CERT Guide to Insider Threats: How to Prevent, Detect, and Respond to Information Technology Crimes (Theft, Sabotage, Fraud). Addison-Wesley, Boston (2012)
17. Chakraborty, S., Ray, I.: TrustBAC: integrating trust relationships into the RBAC model for access control in open systems. In: Proceedings of the Eleventh ACM Symposium on Access Control Models and Technologies, SACMAT 2006, New York, NY, USA, pp. 49–58. ACM (2006)
18. Chen, L., Crampton, J.: Risk-aware role-based access control. In: Meadows, C., Fernandez-Gago, C. (eds.) STM 2011. LNCS, vol. 7170, pp. 140–156. Springer, Heidelberg (2012). https://doi.org/10.1007/978-3-642-29963-6_11
19. Chen, T., Kammüller, F., Nemli, I., Probst, C.W.: A probabilistic analysis framework for malicious insider threats. In: Tryfonas, T., Askoxylakis, I. (eds.) HAS 2015. LNCS, vol. 9190, pp. 178–189. Springer, Cham (2015). https://doi.org/10.1007/978-3-319-20376-8_16
20. Davidson, E.H., Erwin, D.H.: Gene regulatory networks and the evolution of animal body plans. Science **311**(5762), 796–800 (2006)
21. Davis, M., Putnam, H.: A computing procedure for quantification theory. J. ACM **7**(3), 201–215 (1960)
22. de Moura, L., Bjørner, N.: Z3: an efficient SMT solver. In: Ramakrishnan, C.R., Rehof, J. (eds.) TACAS 2008. LNCS, vol. 4963, pp. 337–340. Springer, Heidelberg (2008). https://doi.org/10.1007/978-3-540-78800-3_24
23. Dimmock, N., Belokosztolszki, A., Eyers, D., Bacon, J., Moody, K.: Using trust and risk in role-based access control policies. In: Proceedings of the Ninth ACM Symposium on Access Control Models and Technologies, SACMAT 2004, New York, NY, USA, pp. 156–162. ACM (2004)
24. Dressler, F.: Self-organized network security facilities based on bio-inspired promoters and inhibitors. In: Dressler, F., Carreras, I. (eds.) Advances in Biologically Inspired Information Systems, pp. 81–98. Springer, Heidelberg (2007). https://doi.org/10.1007/978-3-540-72693-7_5
25. Feng, F., Lin, C., Peng, D., Li, J.: A trust and context based access control model for distributed systems. In: 2008 10th IEEE International Conference on High Performance Computing and Communications, pp. 629–634, September 2008
26. Gheyas, I.A., Abdallah, A.E.: Detection and prediction of insider threats to cyber security: a systematic literature review and meta-analysis. Big Data Anal. **1**(1), 6 (2016)

27. Glasser, J., Lindauer, B. : Bridging the gap: a pragmatic approach to generating insider threat data. In: 2013 IEEE Security and Privacy Workshops, pp. 98–104, May 2013
28. Legg, P.A., Buckley, O., Goldsmith, M., Creese, S.: Automated insider threat detection system using user and role-based profile assessment. IEEE Syst. J. **11**(2), 503–512 (2017)
29. Ma, J., Adi, K., Mejri, M., Logrippo, L.: Risk analysis in access control systems. In: 2010 Eighth International Conference on Privacy, Security and Trust, pp. 160–166, Aug 2010
30. Nissanke, N., Khayat, E.J.: Risk based security analysis of permissions in RBAC. In: WOSIS (2004)
31. Nurse, J.R.C., et al.: Understanding insider threat: a framework for characterising attacks. In: 2014 IEEE Security and Privacy Workshops, pp. 214–228, May 2014
32. Zhang, R., Chen, X., Shi, J., Xu, F., Pu, Y.: Detecting insider threat based on document access behavior analysis. In: Han, W., Huang, Z., Hu, C., Zhang, H., Guo, L. (eds.) APWeb 2014. LNCS, vol. 8710, pp. 376–387. Springer, Cham (2014). https://doi.org/10.1007/978-3-319-11119-3_35
33. Rashid, T., Agrafiotis, I., Nurse, J.R.C.: A new take on detecting insider threats: exploring the use of hidden markov models. In: Proceedings of the 8th ACM CCS International Workshop on Managing Insider Security Threats, MIST 2016, New York, NY, USA, pp. 47–56. ACM (2016)
34. Rauf, U.: A taxonomy of bio-inspired cyber security approaches: existing techniques and future directions. Arab. J. Sci. Eng. **43**, 6693–6708 (2018)
35. Salim, F., Reid, J., Dawson, E., Dulleck, U.: An approach to access control under uncertainty. In: 2011 Sixth International Conference on Availability, Reliability and Security, pp. 1–8, August 2011
36. Song, Y., Salem, M.B., Hershkop, S., Stolfo, S.J.: System level user behavior biometrics using Fisher features and Gaussian mixture models. In: 2013 IEEE Security and Privacy Workshops, pp. 52–59, May 2013
37. Ted, E., et al. Detecting insider threats in a real corporate database of computer usage activity. In: Proceedings of the 19th ACM SIGKDD International Conference on Knowledge Discovery and Data Mining, pp. 1393–1401 (2013)
38. Thomas, L.C., d'Ari, R.: Biological feedback. CRC Press, Boca Raton (1990)

Blinded by Biology: Bio-inspired Tech-Ontologies in Cognitive Brain Sciences

Paola Hernández-Chávez[(✉)]

Center for Philosophy of Science, University of Pittsburgh, Pittsburgh, USA
hcpaola@gmail.com

Abstract. In his pioneering paper on neuromorphic systems, Carver Mead conveyed that: "Biological information-processing systems operate on completely different principles from those with which most engineers are familiar" (Mead 1990: 1629). This paper challenges his assertion. While honoring Mead's exceptional contributions, specific purposes, and correct conclusions, I will use a different line of argumentation. I will make use of a debate on the classification and ordering of natural phenomena to illustrate how background notions of causality permeate particular theories in science, as in the case of cognitive brain sciences. This debate shows that failures in accounting for concrete scientific phenomena more often than not arise from (1) characterizations of the architecture of nature, (2) singular conceptions of causality, or (3) particular scientific theories – and not rather from (4) technology limitations *per se*. I aim to track the basic bio-inspiration and show how it spreads bottom-up throughout (1) to (4), in order to identify where bioinspiration started going wrong, as well as to point out where to intervene for improving technological implementations based on those bio-inspired assumptions.

Keywords: Natural kinds · Cognitive brain sciences · Ontology · BioInspiration · Technology

> *Listen to the technology and find out what it's telling you.*
>
> Carver Mead

1 Introduction

Ontological conceptions are background ideas that pervade the practice of science, technology, and their contrivances. A typical example is that, for a system to be explained scientifically, it must be the kind of thing that admits a mechanical account. Ontological conceptions are ways of framing a problem that we often take for granted. One can go to a psychotherapist to reflect upon the hidden motivations for their behaviors. But we do not go to the ontologist to analyze the assumptions we initially embrace about the structure of the world. Since we do not often reflect upon these assumptions, the consequences of adopting them are not addressed. We can step back to draw attention to them when facing technological implementation difficulties such as how stable systems are in general, or the extent to which humans can track microscopic states.

Classifying natural phenomena into general regularities reflects a human proclivity to optimize our understanding of the world. Classification and categorization are recurring practices throughout the history of science. Metascientific studies of how scientists allocate different systems of classifications go back and forth in physics, biology, and recently in cognitive brain sciences. This is not surprising, given the impact these ones have in terms of technological resources, numerous results, and the public attention it draws. Yet, few analyses have been run regarding the origins of neuroscience classifications and the bio-inspired causal assumptions they involve.

As we will see, the bio-inspired causality assumptions at work in current cognitive brain sciences can be traced back to a causal agent-based model coming mainly from nineteenth-century classical biology – which is very much alive.

The following section gives an account of natural kinds and the multiple realization debates in philosophy. The purpose of discussing these concepts is to state the basic terminology and general background for addressing the problem at hand, as well as to draw awareness to the way ontological assumptions and conceptualizations permeate technological tools for explaining particular natural phenomena. Talking about the organization of nature will allow us to transition from biology to medicine, neurology, and finally to cognitive brain sciences. The third section is a quick overview of cognitive brain sciences' range of classification schemes. The fourth section surveys the roots of common inspiration between biology and cognitive brain sciences. The fifth and final section offers recommendations for improving bio-inspiration.

The main contribution of this paper consists in tracking the inspirational explanatory patterns common to biology and brain science technologies, showing why it is necessary to stop importing causal assumptions from classical biology, and where to intervene to improve technological implementations based on those bio-inspired assumptions. To get to these conclusions, we must start with some basic notions from philosophy.

2   Some Philosophical Background: Natural Kinds and Multiple Realizability

The purpose of this section is to define a basic terminology and the general background ideas needed to address the various ways in which natural phenomena could be described. There is an old philosophical debate regarding our most basic notions of the organization of nature and of causality, known as the natural kinds (NK) debate. If you are among those endorsing the idea that our basic scientific taxonomies depict the exact organization of nature as it really is, a philosopher might claim you are a "universalist," a "realist," or an "essentialist." This side of the debate encompasses the intuition that the labor of science is to group phenomena based on their properties, causal relationships, and governing laws. On the other side is instrumentalism, or pragmatism. This is a slightly looser perspective. Kinds and scientific categories function as useful tools to grouping phenomena only for the sake of providing explanations, predictions, or elaborating reliable inferences.

Both sides have persuasive points. On the one hand, an instrumentalist might say that belief in natural kinds exhibits a (wrong-headed) faith in the order and regularity of nature. On the other, a merely instrumental use of scientific classifications does not guarantee accuracy, since the history of science teaches us that scientific practice can prevail for centuries, even millennia, grounded on spurious categories and utterly wrong beliefs. The shift from geocentrism to heliocentrism is the best exemplar of this case.

There is another way that ontology can inspire our causal assumptions: through multiple realizability (MR). MR is the possibility of achieving the same goal by causally different routes. For example, having two distinct entities (ex. parts of the body, components, neural substrates) performing the same function (ex. mental state, cognitive task, instruction) by different operation modes could count a case for MR.

For twentieth century philosophers of mind, it was almost a truism that psychological states are multiply realizable (Putnam 1967) - that a mental state like feeling pain could be realized in different (physical) ways across species, and even between individuals of the same species. In the same vein, a cognitive state could be accomplished by different brain substrates. This is similar to having two subjects performing the same cognitive task while a brain scanning shows that they elicit different brain activity patterns.

MR has some resemblance to redundancy in biology, a common event where a gene is duplicated within a genome of a complex organism. When a system is interrupted by a backup condition or a compensatory response, for example, redundancy helps to facilitate the central functioning and maintenance of the system. In such cases, two or more genes in the genome of an organism can be performing the same function but the activation in one of them happens to have no effect on the phenotype. There are three types of genetic redundancy: true, generic, and 'almost'. In true redundancy, a subject with a redundant genotype AB happens to be no fitter than a second one who has redundant genes being mutated or knocked out. In generic redundancy, an AB subject could be occasionally fitter. However, in an 'almost' redundancy case, the redundant genotype is slightly fitter than any other genotype where one of the redundant genes has been mutated or knocked out (see Nowak et al. 1997).

Several authors have written against MR. One of its most remarked inconveniences is that, as Kim (1992) noticed decades ago, multiply realizable functional kinds are not projectible –owning a predicate to project properties upon it–, which makes it complicated to nominate them as candidates of scientific kinds. This is, when kind members lack an underlying causal basis for membership in that kind, we cannot make inductive generalizations about the nature of that kind. From here, Kim (1992) conjectured that kinds with different physical realization are distinct kinds - structure independent kinds that do not count as causal kinds - so they are not proper scientific kinds.

It is easy to anticipate why MR is ontologically compromising for scientists. Systematizing experiments with possibly 'multiple' outcomes makes science less defensible since experimental configurations would be difficult to model. In spite of this, it is essential to consider that such an uncomfortable possibility might be the case.

Neglecting NKs to avoid ontological commitments became common ground. Philosophers moved on from arguing over natural kinds and so did scientists at stopping debating the reality of natural kinds to avoid ontological commitments. The discussion settled on a softened, naturalized or more scientific friendly conception of "scientific kinds," the hallmark of which was a profound reliance on prosperous scientific practice. This is, further formulations derived from successful (or productive) scientific practices derived from analyzing how scientists use classifications in practice. A sophisticated approach was Richard Boyd's discussion of homeostatic property clusters (Boyd 1990, 1991, 1999a, 1999b, 2000; Boyd 2003). Boyd contended that nature doesn't neatly divide into well-delimited sorts of things. Instead, groupings might be modified in light of new observations or when inferences fail, since definitions of kinds are *a posteriori* (Boyd 2000, 54). For him, kinds used in sciences are not features of the world but products of our engagement with it. As he puts it: "the theory of natural kinds is about how schemes of classification contribute to the formulation and identification of projectible hypotheses" (Boyd 1999a: 147). This means that NKs are groups of entities that share a cluster of projectable properties sustained by homeostatic causal mechanisms, which are understood as anything that causes a repetitive clustering of properties. Consequently, if we cannot make projections and inductive generalizations of a case – say, if the case is multiply realizable – that case cannot be considered as a NK.

Ereshefsky and Reydon (2015) proposed an influential reformulation of the debate that shifted the focus to scientific practice. There they aimed to track a variety of classificatory practices of successful science. They pursued a NKs account that recognizes diverse scientists' aims when constructing classifications. Yet, they denied that any classification offered by scientists correspond to NKs.

Ereshefsky and Reydon (2015) introduce the idea of a "classificatory program", understood as the part of a discipline that produces a classification. According to them, a classificatory program contains three parts: sorting principles, motivating principles, and classifications. Classifications describing NKs are marked by: internal coherence, empirical testability, and progressiveness. The primary virtue of Ereshefsky and Reydon's (2015) proposal is their lack of commitment to the existence of ontological NKs. They introduced several criteria for what makes a classification of natural kinds valid (its internal coherence, empirical testability, etc.), which shifts the frame for analyzing classifications away from 'are these true natural kinds?' to 'do these natural kinds work in the context of scientific practice?'.

But some have argued that, in their effort to encompass all successful classificatory practices, Ereshefsky and Reydon distort some heuristics and common practices in science. However, scrutinizing such critiques is beyond the purposes of this work. Suffice it to say that the possibility of MR in the context of natural kinds raises the issue of whether natural kinds are ontologically real or just an instrumental/methodological guide. Ideally, kind members should be caused by the real properties of the world, and we should be able to identify and specify the causal mechanisms responsible for the grouping of kind members. Nevertheless, arbitrary classifications may occur.

It is crucial to be aware of the fact that an Essentialist, who by definition is committed to the ontological existence of NKs, in his willingness to avoid the possibility of MR, would need to postulate an *overflowing* amount of kinds – as many as one for almost every new phenomenon (each corresponding to a particular law of Nature). But closed cause-effect events are scarce. The paradox is that for the Essentialist the ontological search for NKs as consisting of a system of neat, simple, and well-organized frames of Nature loses its original appeal: that of unraveling the laws of an engineered Nature. By contrast, since an Instrumentalist only needs to describe Nature in such a way that it works for science, he would readily allow the incursion of MR. He would then indirectly promote a more *parsimonious* amount of causes, admitting only those coming from the most basic physics, for example. Ironically, for an Instrumentalist Nature would be less chaotic than for an Essentialist.

The take-home lesson from this survey on NK and MR is the importance of being aware of the ontological commitments brought by our intuitions about how the world and natural phenomena are arrayed. If your intuition tells you, for example, that the human mind is modular, and that the most basic operations of it are computational, this will permeate the technological implementation of your experiments, perhaps by means of scaling up the power of computational outcomes to make them the most fundamental part of brain cognitive operations – regardless of whatever related processes remain unaccounted for. As another example: if you have sympathy for NKs, it is more likely that you start a paradigm that correlates a brain region to remembering as a broad category, then looking for additional brain regions for particular tasks, associating particular regions to each thing we humans do. On the contrary, if you accept MR and instrumentalism in principle, it would be plausible to claim that there is nothing essential to remembering, so you will find no troubles transiting from that nothingness, to game theory, to modeling accordingly. Overall, starting from scratch might look like a challenging and complex outset, but it will leave you the option of using any explanatory theory or methods that allows you to understand and predict phenomena. On the contrary, starting from within a NKs framework might lead you to a very chaotic multiplication of categories. This option seems to be less defensible, since there has been no long-lasting scientific system of categories – or maybe there are just a very few ones. This is most likely because, in the effort to control, predict and discover new natural phenomena, science constantly invents emerging classification systems.

With this philosophic background in mind, we can turn to a comparison of explanatory patterns in brain sciences and in biology. I will pursue this comparison by first elaborating on the landscape of explanatory frameworks in the brain sciences.

3   Cognitive Brain Sciences Landscapes and Classifications

Cognitive accounts of the brain are abundant. Diverse descriptions are found depending on the different collection of components, hierarchies, and type of interactions between them. Agreement about which are the correct set of components of cognitive processes in the brain remains far away.

Many different lists of cognitive brain sciences components might be found. For the sake of space and with the purpose of illustrating the point, let us grant that there exist only two basic models. The most basic level of brain cognition, what I will call the "atomistic model," would include the following components:

a molecular scale
ion channels + neurotransmitters
synapses
neurons
neural networks
neural systems

On the other hand, we could elaborate another list for the "cognitive perspective model." This would include the following components:

brains
a physical body (embodied experience)
subjects
interacting subjects
culture / environmental elements (extended cognition)

From this perspective, the cognitive brain and its components would be literally extended in multifactorial ways.

Why is integration among this dualistic simplified version of brain and cognition still so challenging for theorists, engineers, and scientists in general? Following the discussion in section two, I suspect this is because different models imply a different ontological array of natural components.

The ontological components comprising a scientist's favorite system will determine the level to which each one pays more attention. Among the possible elements of composition that a scientist could emphasize are: the scale of interest (a fine-grained one, or even a nanotechnological perspective, as opposed to big data collaborative projects)[1], the multiscale (a computer simulation of interactions vs. engineering a whole neural systems), the method (ex. reverse engineering as a tool for emulating micro-structures interactions), the elements to simulate (ex. mimicking synaptic transmission arrays), the spatial and temporal resolution, the logic of the behavior (decision making vs. game theoretic), or the underlying rules (representational vs. connectionist responses; or connectionist vs. modular responses), among others.

---

[1] Notable approaches to the study of the brain are the Human Connectome Project (USA) and the Human Brain Project Initiatives (Europe). Despite their refractory differences, both concur that the fundamental puzzles in Neuroscience are:

- Deciphering the primary language of the brain
- Understanding the rules governing how neurons organize into circuits;
- Understanding how the brain communicates information from one region to another, and which circuits to use in a given situation;
- Understanding the relation between brain circuits, genes, and behavior;
- Developing new techniques for analyzing and observing brain function.
- Disentangling the essential elements of neural computation.

Having sketched the connection between ontology and cognitive brain sciences, and the relevance of this connection, the following section addresses the common explanatory patterns that cognitive brain sciences share with biology and where these patterns come from.

4 Blinded by Biology

The search for ideological principles guiding cognitive brain sciences refers us to the roots of scientific medical thinking, itself so much inspired by a notion of causality coming from nineteenth-century biology. There, an atomistic logic pattern prevails: a single agent 'S' causes infection 'I'. Until recently, biology had focused on genes as the definers of structure and function, where the agent is understood as only a physical realization of the genetic program, without regard for the dynamics, interactions and breakout patterns of genes.

Causal agent-based explanatory models in biology and medicine enjoyed a glorious era in the 19th centuries. Illnesses such as cholera are great examples of successful causal pathogen identification: bacterium *vibrio cholera* causes cholera, *treponema pallidum* causes syphilis, *H1N1* causes swine flu. Under this line of thought, once you identify the pathogen, it is possible to track its dynamics and predict its course of action, so that subsequent prognosis and models of intervention arise.

The same atomistic causal explanatory practices have lasted right up to the present day. Approaches to the study of cognition and its brain substrates use tools such as: (1) the review of empirical evidence in subjects suffering from brain damage or selective cognitive impairments as isolated components; (2) the study of selective deficits in atypical subjects, i.e., with a neurodevelopmental disorder or a brain damage, also taken as units; and (3) the use of neuroimaging techniques (PET, fMRI, etc.) aimed to register brain activity related to the performance of very particular cognitive tasks – again, taken as units. These tools or techniques embody a linear causal pattern coming from classical biology. These approaches could be seen as trying to test counterfactual situations – like with those who suffer from neurological dysfunctions, being the logic of counterfactuals the same kind of linear causal logic.

The operating logic of classical biology proved to be quite successful for over three centuries. But, for several reasons, it became indisputable that the gene-reductionist paradigm in biology was not a good way to start accounting for the function of a vast number of illnesses. As those authors claim, classical questions in biology as to what life is, what nature is, what an organism is, how they are organized, what their regulating functions are, etc., were considered metaphysical issues (Laubichler 2000; Cornish-Bowden 2006; Nicholson 2014). That is why favoring gene-centered accounts received so much attention, provided the extensive deployment of empirical tools they invoked –leaving the methapysical voids aside–.

Similarly, as far as cognitive brain science goes, that logic might not be the best route to approach the understanding of the brain and cognitive phenomena provided the conceptual gaps between a brain activation elicited by a cognitive task, the problems with the basic framework assumptions in regards to brain and cognition, the instrumental difficulties elicited by current brain scanning techniques, among others (see Hernández Chávez 2019, forthcoming). But how to overcome this biology-inspired atomistic causality?

5 Improving Bio-inspiration

What kind of biology could cognitive brain scientists look to for inspiration? I argue that systems biology offers a better way forward, far beyond nineteenth-century biology, thus representing a better source of inspiration.

A shift took place in biology when some biologists stopped focusing only on genes. As some authors noticed, the term 'organism' almost disappeared from mainstream literature as a fundamental explanatory concept in biology (Webster and Goodwin 1982; Laubichler 2000; Huneman and Wolfe 2010; Cornish-Bowden 2006; Nicholson 2014). The shift accomplished by systems biology focused on the following facts: genes live within an organism, and they form complex systems; organisms, not genes, are the agents of evolution; phenotypic plasticity is the rule (genotypes generate different phenotypes depending on the environmental circumstances); phenotypic innovations can be genetically inherited; and, organisms are heterogeneous, and their dynamicity and variability more precisely characterize them. This additionally promoted niche construction approaches, which underlined how organisms modify their environment and also inherit ecological changes (for a quick overview, see Laland et al. 2016).

Given the relationships depicted in Fig. 1, one would expect innovations in systems biology to penetrate reciprocally into cognitive brain sciences. Yet, cognitive brain sciences accounts have not been updated to take into account systems biology conceptualizations. This is my plea for cognitive brain scientists to do so. A systems approach for cognitive brain sciences would promote putting the brain back into the center of the discussion, thus modeling brain operation and functioning as a complex phenomenon.

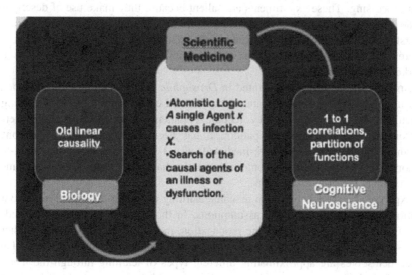

Fig. 1. Represents the possible causal route going from assumptions in classical biology to the atomistic logic of scientific medicine, and then to cognitive brain sciences. As the figure tries to show, the mindset still operating in current cognitive brain sciences was most likely imported from nineteen-century biology.

Noble's (2012) modeling of heart cells is an interesting case of going beyond a simplistic, classical bio-inspired causality framework. Noble found that as it occurs in all functions that require cellular structural inheritance as well as genome inheritance, there is not a program for cardiac rhythm in the genome. Those findings led him to accept that "we cannot yet characterize all the relevant concentrations of transcription factors and epigenetic influences. [so that] It is ignorance of all those forms of downward causation that is impeding progress" (Noble 2012: 60). As this case shows, a move away from classical bio-inspiration concerning the architecture of natural organization brought him to acknowledge that there is no privileged scale at which biological functions are determined. Why this is a good thing? What was gained from the recognition that there is no privileged scale at which biological functions similarly applies as the upshot for practitioners of cognitive scientists as well.

Another example of overcoming simplistic causality is Marr and Hildreth's (1980) model of edge detection based on intensity changes occurring within an image over a range of scales, where each of them are detected separately at different frames. They discovered that intensity changes tend to emerge from surface discontinuities, reflectance, or illumination boundaries. The most notable characteristic is that they all share the property of being spatially localized, so that there are crossing segments coming from different but non-independent channels – so much so that the operating rules can be deduced from a sketch-like combination of images. The relevant payout of this theory is that it succeeded in advancing several psychophysical discoveries as to how oriented segments are formed.

Some additional cases from studies of insects might be of help for building improved causal frameworks. Recent experiments demonstrate insects' capacity to learn and memorize complex visual arrays that eventually carrying with a modular brain processing. These experiments are salient because they make use of descriptions of the content learned through visual stimuli in combination with a generalized understanding of learned and unlearned routines such that the memory of 'what' and 'where' are stored differently than what is supposed in traditional accounts (see Sztarker and Tomsic 2011).

Similarly, it has been documented in *Drosophila* discrimination and remembering of visual landmarks that select patterns as size and color that are stored according to particular parameter values. They recognize patterns independently of the retinal position and acquisition of the pattern. It has been shown that they also contain something similar to a network-mediated visual pattern recognition. Short-term memory traces of elevation and contour orientation have also been documented, among other findings (see Liu et al. 2006).

Experimental models based on machine learning also offer additional routes out of traditional and simplistic causal assumptions. In those cases, models are carried on trained neural networks to study the propagation, modes of learning, speed, gradients, dimensional patterns, algorithms, processing, and many other features. Neural networks themselves are approximating different types of learning through the manipulation of variables like propagation and speed. More recently, deep learning is becoming quite influential in bringing together engineering and reinforcement learning. Within those frameworks, complex control problems can be successfully explained not by invoking basic, general operation rules of causality, but rather stochastic, nonlinear, and autonomous dynamics.[2]

Many different research programs now provide improved perspectives for systems dynamics models by going beyond traditional approaches so to include: reinforcement of learning, positive reinforcement, distributed agency, cooperation in game theories, decision making, action control, systems resolutions, alignment of utilities, to name a few. Modeling brain sciences after these efforts would be as fruitful. Those are thus better sources of bio-inspiration when elaborating models for cognitive brain sciences that promote new emerging assemblies and properties.

6 Concluding Remarks

In this work, I highlighted how theoretical models are often more powerful than we think, insofar as they influence explanations and technological models of phenomena. To a significant extent, cognitive brain science in particular is bio-inspired by causal assumptions in classical nineteenth-century biology. I argued that an approach inspired by systems biology would be more suitable for understanding human brain dynamics and cognition. This would output back integration, regulation, and organization of the living phenomenon, innovation, among others. In other words, approaches inspired by systems biology would put the focus back on investigating the integration, regulation, and organization of living phenomena *as a whole* –instead of treating organisms as hierarchical assemblies of generic basic components.

I made use of the natural kinds and multiple realizability debate on the ordering of natural phenomena (i.e. discussions of whether nature is a messy place, or if it is well organized; whether phenomena are subject to regularities, etc.) to illustrate how background notions of causality permeate theories in science (ex. atomistic/agent based ones, complex phenomena mindset, indeterminism). In the case of cognitive brain sciences, I examined those background notions of causality evident in different initiatives that focus on learning and connectionist patterns, modular explanations, and input-output processing outsets. I argued that, more often than not, the failures of these initiatives to account for concrete scientific phenomena most likely arise from (1) characterizations of the architecture of natural phenomena, (2) singular conceptions of causality, and (3) particular scientific theories, and not rather from (4) technology limitations *per se*. So I tracked the basic inspiration for these background notions of causality from classical biology, and how this inspiration spread bottom-up throughout (1) to (4). Thus, it suggested it is possible to identify where to intervene to improve technological implementations.

In general, humans and other animals master perceiving the world as a set of uniform and organized structures. As we improve, some aspects become less meaningful. A substantial number of reductions take place – such as ignoring tridimensional

---

[2] See for example, Ng et al. (2006).

arrays, color palette nuances, minimizing complexities, making complex systems to appear as binary, or any form of dualism – which promotes the blatant reduction of representations. To that extent, brain cognitive operations are "crafted functional abstractions" (Cauwenberghs 2013: 15513). These strategies are not necessarily mistaken, since they promote efficient learning of routines and fast processing, among other virtues. Yet, it is paramount to remain aware of those simplifications by reassessing the levels of simplification/complexity involved and the overall size of the full scale, so as to know: where and how a model is scaling up, what exactly is being computed or simulated, which level of abstraction is being taking for granted and (more importantly) at what cost. Such is the importance of analyzing seriously the quality of bio-inspiration operative models in the cognitive brain sciences.

Being aware of our most basic assumptions going from (1) to (4) wards off contamination in technology-based explanatory models in science, as in the case of brain and cognition functioning and dynamics. We need accuracy in technologies, but we also need real awareness of our ontological commitments in order to have technologies that are well-matched to the explanatory task at hand.

References

Boyd, R.N.: Realism, approximate truth, and philosophical method. In: Savage, C.W. (ed.) Scientific Theories, pp. 355–391. University of Minnesota Press, Minneapolis (1990)

Boyd, R.N.: Realism, anti-foundationalism and the enthusiasm for natural kinds. Philos. Stud. 61 (1–2), 127–148 (1991)

Boyd, R.N.: Kinds, complexity, and multiple realization. Philos. Stud. 95(1–2), 67–98 (1999a)

Boyd, R.N.: Homeostasis, species, and higher taxa. In: Wilson, R.A. (ed.) Species: New Interdisciplinary Essays, pp. 141–185. MIT Press, Cambridge (1999b)

Boyd, R.N.: Kinds as the "Workmanship of Men": realism, constructivism, and natural kinds. In: Nida-Rümelin, J. (ed.) Rationalität, Realismus, Revision: Vorträge des 3. Internationalen Kongresses der Gesellschaft für Analytische Philosophie, pp. 52–89. De Gruyter, Berlín (2000)

Boyd, R.N.: Finite beings, finite goods: the semantics, metaphysics and ethics of naturalist consequentialism, Parts I & II. Philos. Phenomenol. Res. LXVI, 505–553 & LXVII, 24–47 (2003)

Cauwenberghs, G.: Reverse engineering the cognitive brain. PNAS, 110(39), 15512–15513 (2013)

Cornish-Bowden, A.: Putting the systems back into systems biology. Perspect. Biol. Med. Autumn 49(4), 475–489 (2006)

Putnam, H.: Psychological predicates. In: Capitan, W.H., Merrill, D.D. (eds.) Art, Mind, and Religion, pp. 37–48. University of Pittsburgh Press, Pittsburgh (1967)

Ereshefsky, M., Reydon, T.A.C.: Scientific kinds. Philos. Stud. 172(4), 969–986 (2015)

Hernández Chávez, P.: Disentangling Cognitive Dysfunctions: When Typing Goes Wrong, (2019, forthcoming)

Huneman, P., Wolfe, C.: The concept of organism: historical philosophical, scientific perspectives. Hist. Philos. Life Sci. 32(2–3), 147–154 (2010)

Kim, J.: Multiple realization and the metaphysics of reduction. Philos. Phenomenol. Res. 52, 1–26 (1992)

Laland, K., Matthews, B., Feldman M.: An introduction to niche construction theory. Evol. Ecol. **30**, 191–202 (2016)

Laubichler, M.D.: The organism is dead. long live the organism! Perspect. Sci. **8**(3), 286–315 (2000)

Liu, G., et al.: Distinct memory traces for two visual features in the *Drosophila* brain. Nature **439**, 551–556 (2006)

Marr, D., Hildreth, E.: Theory of edge detection. Proc. R. Soc. Lond. B Biol. Sci. **207**(1167), 187–217 (1980)

Mead, C.: Neuromorphic electronic systems. Proc. IEEE **78**(10), 1629–1636 (1990)

Ng, A.Y., et al.: Autonomous inverted helicopter flight via reinforcement learning. In: Ang, Marcelo H., Khatib, O. (eds.) Experimental Robotics IX. STAR, vol. 21, pp. 363–372. Springer, Heidelberg (2006). https://doi.org/10.1007/11552246_35

Nicholson, D.J.: The return of the organism as a fundamental explanatory concept in biology. Philos. Compass **9**, 347–359 (2014). https://doi.org/10.1111/phc3.12128

Noble, D.: A theory of biological relativity: no privileged level of causation. Interface Focus **2**, 55–64 (2012)

Nowak, M.A., Boerlijst, M.C., Cooke, J., Smith, J.M.: Evolution of genetic redundancy. Nature **388**, 167–171 (1997)

Sztarker, J., Tomsic, D.: Brain modularity in arthropods: individual neurons that support "what" but not "where" memories. J. Neurosci. **31**(22), 8175–8180 (2011). https://doi.org/10.1523/JNEUROSCI.6029-10.2011

Webster, G., Goodwin, B.C.: The origin of species: a structuralist approach. J. Soc. Biol. Struct. **5**(1), 15–47 (1982). https://doi.org/10.1016/S0140-1750(82)91390-2

A Distribution Control of Weight Vector Set for Multi-objective Evolutionary Algorithms

Tomoaki Takagi[✉], Keiki Takadama, and Hiroyuki Sato

The University of Electro-Communications, 1-5-1 Chofugaoka,
Chofu, Tokyo 182-8585, Japan
tomtkg@uec.ac.jp, keiki@inf.uec.ac.jp, sato@hc.uec.ac.jp

Abstract. For solving multi-objective optimization problems with evolutionary algorithms, the decomposing the Pareto front by using a set of weight vectors is a promising approach. Although an appropriate distribution of weight vectors depends on the Pareto front shape, the uniformly distributed weight vector set is generally employed since the shape is unknown before the search. This work proposes a simple way to control the weight vector distribution appropriate for several Pareto front shapes. The proposed approach changes the distribution of the weight vector set based on the intermediate objective vector in the objective space. A user-defined parameter determines the intermediate objective vector in the static method, and the objective values of the obtained solutions dynamically determine the intermediate objective vector in the dynamic method. In this work, we focus on MOEA/D as a representative decomposition-based multi-objective evolutionary algorithm and apply the proposed static and dynamic methods for it. The experimental results on WFG test problems with different Pareto front shapes show that the proposed static and dynamic methods improve the uniformity of the obtained solutions for several Pareto front shapes and the dynamic method can find an appropriate intermediate objective vector for each Pareto front shape.

Keywords: Multi-objective optimization · Evolutionary computation

1 Introduction

Real-world optimization problems often involve multiple conflicting objectives. These problems are said to be multi-objective optimization problems. The aim of multi-objective optimization is to acquire a set of solutions approximating the Pareto front, the optimal trade-off among conflicting objectives. So far, evolutionary algorithms have been intensively studied for solving multi-objective optimization problems [1,2]. As an evolutionary algorithm solving multi-objective problems, we focus on MOEA/D which decomposes the Pareto front in the

A. Compagnoni et al. (Eds.): BICT 2019, LNICST 289, pp. 70–80, 2019.
https://doi.org/10.1007/978-3-030-24202-2_6

objective space [3]. MOEA/D simultaneously optimize a number of scalarizing functions with different weight vectors and tries to approximate the Pareto front by the obtained solutions paired with weight vectors. MOEA/D needs to prepare the set of weight vectors before the search. Since each weight vector determines an approximation part of the Pareto front, the distribution of weight vectors strongly affects the distribution of the obtained solutions in the objective space. The appropriate distribution of weight vectors depends on the shape of the Pareto front. However, the shape of the Pareto front is generally unknown before the search. The conventional MOEA/D uses a uniformly distributed weight vectors based on the simplex-lattice design method. Although several methods arranging the weight vector set have been studied recently [4–7], we are investigating another simple way to re-arrange the weight vector set.

In this work, we propose an approach to re-arrange the weight vector set based on the intermediate objective value in the population. The approach is employed in the two proposed methods. The first one is the static method which determines the intermediate objective value by using an user-defined parameter. The another one is the dynamic method which determines the intermediate objective value by the objective values of the obtained solutions during the search. The effects of the two proposed methods are verified on tWFG4 problems [8] with different Pareto front shapes.

2   Evolutionary Multi-objective Optimization

A multi-objective optimization problem is defined as follows:

$$\text{Minimize} \quad \boldsymbol{f}(\boldsymbol{x}) = (f_1(\boldsymbol{x}), f_2(\boldsymbol{x}), \ldots, f_m(\boldsymbol{x})) \tag{1}$$

Since these minimizations of m kinds of objective functions f_i $(i = 1, 2, \ldots, m)$ often conflict, there is no single optimal solution x which minimizes all objective values simultaneously. The optimal trade-off among objectives arises instead, and it is said to be the Pareto front. Therefore, the task of multi-objective optimization is to acquire a solution set approximating the Pareto front. Approximation qualities of the Pareto front can be evaluated by the convergence of solutions toward the Pareto front, the spread of them to cover the entire Pareto front, and the distribution uniformity of them in the objective space.

Although there are several approaches to address multi-objective optimization, population-based evolutionary algorithms have an advantage that a set of solutions approximating the Pareto front can be obtained in a single run. In this work, we focus on MOEA/D [3] as a representative evolutionary algorithm for solving multi-objective optimization problems.

3   MOEA/D

MOEA/D specifies N approximating parts of the Pareto front with a weight vector set $\mathcal{L} = (\boldsymbol{\lambda}^1, \boldsymbol{\lambda}^2, \ldots, \boldsymbol{\lambda}^N)$. Each weight vector $\boldsymbol{\lambda}^i$ has m kinds of elements $(\lambda_1^i, \lambda_2^i, \ldots, \lambda_m^i)$. The decomposition parameter H and the number of objectives m are used to generate the weight vector set $\mathcal{L}$. Each element of $\boldsymbol{\lambda}^i = (\lambda_1^i, \lambda_2^i, \ldots, \lambda_m^i)$ is one of $\{0/H, 1/H, \ldots, H/H\}$, and all $N = C_{H+m-1}^{m-1}$ kinds of weight vectors satisfying $\sum_{j=1}^{m} \lambda_j^i = 1.0$ are employed in the search. These weight vectors are uniformly distributed on the m-dimensional hyperplane, and it is said to be the simplex-lattice design.

Since each weight vector $\boldsymbol{\lambda}^i$ is paired with one solution $\boldsymbol{x}^i$, the size of the population $\mathcal{P} = \{\boldsymbol{x}^1, \boldsymbol{x}^2, \ldots, \boldsymbol{x}^N\}$ becomes N which is equivalent to the size of the weight vector set $\mathcal{L}$. MOEA/D compares solutions based on their scalarizing function values calculated with their objective function vectors and a weight vector. For each weight vector, MOEA/D maintains the best solution with the minimum scalarizing function value in the population. Although there are several scalarizing functions, in this work, we use the reciprocal weighted Tchebycheff function (rTCH) [9]. The rTCH scalarizing function is formulated as

$$\text{Minimize} \quad g(\boldsymbol{x}|\boldsymbol{\lambda}^i) = \max_{1 \leq j \leq m} \left| \frac{f_j(\boldsymbol{x}) - z_j}{\lambda_j^i} \right|. \tag{2}$$

where, z_j $(j = 1, 2, \ldots, m)$ is an element of the obtained ideal point $\boldsymbol{z}$. z_j is set to the minimum (best) objective function value f_j found during the search.

4   Issue Focus: Weight Vector Distribution

The distribution of the obtained solutions in the objective space is affected by the distribution of the weight vectors. Since the Pareto front shape is generally unknown before the search, the conventional MOEA/D uses the weight vector set uniformly distributed on the m-dimensional hyperplane. However, the weight distribution is appropriate only when the shape of the Pareto front is the hyperplane. For other Pareto front shapes such as convex and concave shapes, their appropriate distributions of weight vectors are different. If the distribution of the weight vectors can be changed according to the Pareto front shape, the distribution uniformity of the obtained solutions can be improved.

Several approaches varying the distribution of weight vectors have been studied so far [4–7]. Jiang et al. proposed a uniformly distributed weight vectors based on the mixture uniform design method [4]. This method changes the curvature of the hyperplane on the weight vectors. Deb et al. proposed a method to change weight vector (reference point) distribution by deleting and adding weight vectors during the search [5]. On the other hand, Hamada et al. proposed adding and re-arranging mechanism of the weight vectors during the search [6,7]. In this work, we propose an alternative way to change the weight vector distribution.

5   Proposed Method: Weight Vector Distribution Control Based on Intermediate Objective Value

The proposed method varies the conventional weight vector set $\mathcal{L}$ to $\mathcal{L}' = \{\boldsymbol{\lambda}^{1'}, \boldsymbol{\lambda}^{2'}, \ldots, \boldsymbol{\lambda}^{N'}\}$ by using the intermediate objective value p. p is a real value in the range $[0, 1]$.

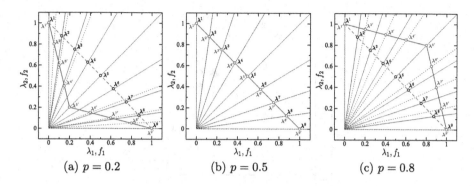

Fig. 1. Weight vector distributions by varying intermediate objective value p

In this work, we proposed two methods. One is the static approach using p as a static user-defined parameter, and another one is the dynamic approach determining p based on the objective values in the population during the search.

5.1 Static Approach

In the static approach, we transform each weight vector $\boldsymbol{\lambda}^i = (\lambda^i_1, \lambda^i_2, \dots, \lambda^i_m)$ into $\boldsymbol{\lambda}^{i'} = (\lambda^{i'}_1, \lambda^{i'}_2, \dots, \lambda^{i'}_m)$ with an user-defined intermediate objective value p in the range $[0, 1]$. Each element is transformed by

$$\lambda^{i'}_j = \begin{cases} \lambda^i_j \cdot p \cdot m, & \text{if } \lambda^i_j \leq \frac{1}{m}, \\ 1 - (1 - \lambda^i_j) \cdot \frac{1-p}{m-1} \cdot m, & \text{otherwise.} \end{cases} \tag{3}$$

Figure 1 shows three examples of transformed weight vector sets with different p on $m = 2$ objective space. We can see that the weight distribution with $p = 1/m = 0.5$ is the equivalent to the conventional weight distribution. We can see that the weights with $p < 0.5$ get close to the origin point. This distribution is appropriate for the convex Pareto front. Also, we can see that the weights with $p > 0.5$ get away from the origin point. This distribution is appropriate for the concave Pareto front.

5.2 Dynamic Approach

The dynamic approach determines the intermediate value p based on the objective values of solutions in the population during the search.

In each generation, the proposed dynamic approach calculates the intermediate objective value p by

$$p = \frac{1}{\sqrt{m}} \cdot d_1(\boldsymbol{x}_L), \tag{4}$$

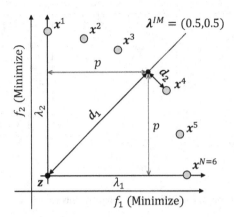

Fig. 2. Calculation of the intermediate value p in the proposed dynamic approach

where,

$$x_L = \arg\min_{x \in \{x^1, x^2, \ldots, x^N\}} d_2(x) \tag{5}$$

$$d_1(x) = \frac{\|(f(x) - z)^T \lambda^{IM})\|}{\|\lambda^{IM}\|}, \tag{6}$$

$$d_2(x) = \left\| f(x) - \left(z - d_1(x) \frac{\lambda^{IM}}{\|\lambda^{IM}\|} \right) \right\|. \tag{7}$$

The two distances d_1 and d_2 are employed from the concept of the penalty-based boundary intersection (PBI) [3]. $\lambda^{IM} = (1/m, 1/m, \cdots, 1/m)$ is the intermediate weight vector specifying the center of the Pareto front.

In Eq. (4), first we find the landmark solution x_L with the minimum distance d_2 among the all solutions $\mathcal{P} = \{x^1, x^2, \ldots, x^N\}$. Figure 2 shows an example on an $m = 2$ objective problem. In this example, the landmark solution is x^4 since it has the minimum distance d_2 to the intermediate weight vector $\lambda^{IM} = (0.5, 0.5)$. In Eq. (4), next we calculate the distance d_1 of the landmark solution. Then, we calculate p as the side length of a hypercube which the length of the diagonal line is d_1. In the example of Fig. 2, p is the side length of a square which the length of the diagonal line is d_1 of x^4. In the proposed dynamic approach, the intermediate objective value p is repeatedly updated every generation. That is, the weight distribution is repeatedly changed every generation.

5.3   Expected Effects

For a convex Pareto front, we can expect the weight distribution with $p < 0.5$ obtains more uniformly distributed solutions than the conventional weight vectors. For a concave Pareto front, we can expect the weight distribution with $p > 0.5$ obtains more uniformly distributed solutions than the conventional weight vectors. In the case of the static approach, we need to set a p value

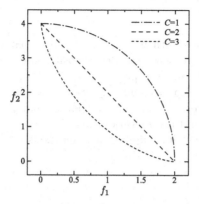

Fig. 3. Pareto fronts with different C values

before the search. To find an appropriate p, we need the parameter tuning of p by repeating run of the algorithm with different p. On the other hand, in the dynamic approach, we can expect to obtain more uniformly distributed solutions than the conventional weights from the convex to the concave Pareto front while searching an appropriate intermediate objective value p during the search.

6   Experimental Setting

6.1   Test Problems

To verify the effectiveness of the proposed methods, in this work, we use tWFG4 [8] problems extended from WFG4 [10] problems. tWFG4 is formulated by

$$f_j(\boldsymbol{x}) = 2j \cdot \left(\frac{f_j^{WFG4}(\boldsymbol{x})}{2j} \right)^C \quad (j = 1, 2, \dots, m), \tag{8}$$

where, f_j^{WFG4} is j-th objective function of WFG4 problem [10], C is a parameter specifying the Pareto front shape. tWFG4 with $C = 2$ has a liner Pareto front. tWFG4 with $C < 2$ has a concave Pareto front. tWFG4 with $C > 2$ has a convex Pareto front. In this work, we use tWFG4 problems with $C = \{1, 2, 3\}$. Figure 3 shows three Pareto fronts with different C.

f_j^{WFG4} $(j = 1, 2, \dots, m)$ involves two problem parameters. First one is the distance parameter L to control the convergence difficulty. Another one is the spread parameter K to control the spread difficulty. To evaluate the uniformity of the obtained solutions, in this work, we respectively set them $K = 1 + (m-1) - 1$ mod $(m - 1)$ and $L = 1$. Consequently, the number of variables becomes $n = K + L = \{3, 3, 4\}$ for $m = \{2, 3, 4\}$. That is, the uniformity of the solutions strongly affect the search performance.

6.2 Parameters

We use tWFG4 problems with $m = \{2, 3, 4\}$ objectives and $C = \{1, 2, 3\}$. For $m = \{2, 3, 4\}$ objective problems, the decomposition parameters are respectively set to $H = \{200, 19, 9\}$, and the population sizes become $N = \{201, 210, 220\}$ in MOEA/D. Also, we employ commonly used SBX with the distribution index $\eta_c = 20$ and the crossover ratio 0.8 and the polynomial mutation with the distribution index $\eta_m = 20$ and the mutation ratio $1/n$. Also, the neighborhood size is set to $T = 20$, and the total number of generations is set to 3,000.

6.3 Metric

To evaluate the obtained solutions, we employ Hypervolume (HV) [11]. HV is a volume determined by the obtained solutions and the reference point r in the objective space. The convergence of solutions toward the Pareto front, the spread of solutions in the objective space, and the uniformity of solutions in the objectives space affects to the HV value. Before we calculate HV, we each normalize objective value as $f'_j(x) = f_j(x)/2j$ $(j = 1, 2, \ldots, m)$. Then we calculate HV with the reference point $r = \{1.1, 1.1, \ldots, 1.1\}$. In this work, we compare the median HV of all independent 31 runs.

7   Experimental Results and Discussion

7.1 Search Performance HV

Figures 4, 5 and 6 show results of HV on problems with different Pareto front shapes $C = \{1, 2, 3\}$ and number of objectives $m = \{2, 3, 4\}$, respectively. In each graph, the horizontal axis indicates the intermediate objective value p of the static approach. The black vertical line on $p = 1/m$ indicates the conventional weight distribution based on the simplex-lattice design. The black horizontal line indicates the HV value obtained by the conventional weight distribution. Also, the red horizontal line indicates the HV value of the proposed dynamic approach. The red vertical line indicates the intermediate objective value p of the proposed dynamic approach at the final generation. The red dot line with makers indicates HV values of the proposed static approach. In all graphs, the error bars are the maximum and the minimum HV values among 31 runs.

First, from Fig. 4 on $m = 2$ objective problems with a plane Pareto front ($C = 2$), we can see that the proposed static approach achieves the highest HV when we use $p^* = 0.5$ which is the equivalent to the conventional weight vector set by simplex-lattice design. We can see that the proposed dynamic approach shows HV value comparable with the one obtained by the proposed static approach with $p^* = 0.5$.

Next, form the results on $m = 2$ objective problems with a concave Pareto front ($C = 1$), we can see that the proposed static approach achieves the highest HV when we use $p^* = 0.8$ which the transformed weight vectors get away from the origin point. Also, we can see that HV achieved by the static approach with

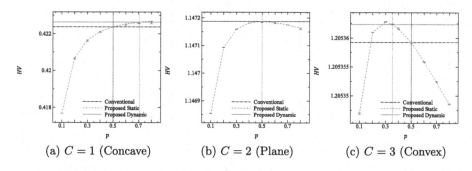

Fig. 4. Relationship between intermediate objective values p and HV in 2 objectives (Color figure online)

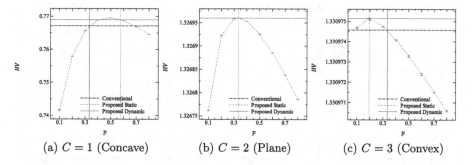

Fig. 5. Relationship between intermediate objective values p and HV in 3 objectives (Color figure online)

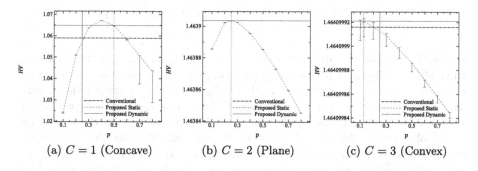

Fig. 6. Relationship between intermediate objective values p and HV in 4 objectives (Color figure online)

$p^* = 0.8$ is higher than HV obtained by the conventional weights with $p = 0.5$. This result reveals that the proposed re-arrangement of weight vectors improves HV value. Also, we can see that the proposed dynamic approach shows HV value close to HV values obtained by the static approaches with $p = \{0.7, 0.8\}$. The static approach has to determine an intermediate objective value p before the

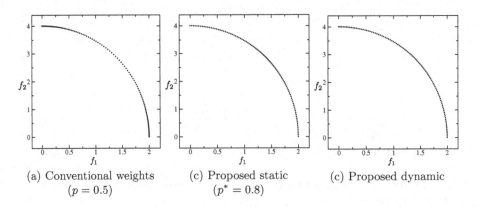

(a) Conventional weights   (c) Proposed static   (c) Proposed dynamic
     ($p = 0.5$)   ($p^* = 0.8$)

Fig. 7. Obtained solutions on the problem a concave Pareto front ($C = 1$)

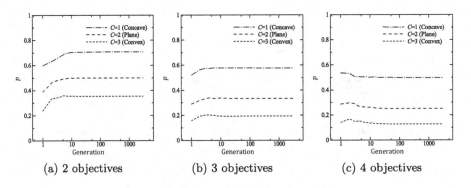

(a) 2 objectives   (b) 3 objectives   (c) 4 objectives

Fig. 8. Transition of the dynamically determined intermediate objective value p

search. To find out the appropriate p^*, we need to repeatedly run the algorithm while slightly changing p. On the other hand, the proposed dynamic approach searches while adjusting the intermediate objective value p during the search. Therefore, the result reveals that the proposed dynamic approach achieves higher HV than the conventional method in a single run. Next, from the results on $m = 2$ objective problems with a convex Pareto front ($C = 3$), we can see that the proposed static approach achieves the highest HV when we use $p^* = 0.3$ which the weight vectors get close to the origin point. Thus, the optimal p^* maximizing HV decreases by increasing C and changing the Pareto front from a concave to a convex. Also, although the proposed dynamic approach cannot achieve the maximum HV obtained by the static approach with the optimal p^*, the proposed dynamic approach achieves higher HV than the conventional method in a single run. On $m = \{3, 4\}$ objective problems, we can see that the similar tendency observed on the $m = 2$ objective problem.

7.2 Obtained Solution Set

Figure 7 shows the obtained solution sets at the final generation on $m = 2$ objective problem with a concave Pareto front ($C = 1$). From the results obtained by the conventional weights, we can see that the approximation granularity around the edges of the Pareto front is high but one in the center of the Pareto front is low. On the other hand, we can see that two proposed methods obtain more uniformly distributed solutions than the conventional weights.

7.3 Transition of p in Dynamic Approach

Figure 8 shows transitions of the intermediate objective value p determined by the proposed dynamic approach. From the result, we can see that p values converge around ten generation on each of the problems with different Pareto front shapes C and the number of objectives. These stable transition of p reveals the stable adaptation of the weight vector distribution by the proposed dynamic approach.

8 Conclusions

To improve the uniformity of obtained solutions on problems with different Pareto front shapes, in this work, we proposed a method to control weight vector distribution based on the intermediate objective value. Experimental results showed that the proposed static approach improves the approximation performance on problems with a convex, a plane, and a concave Pareto fronts by setting an appropriate intermediate objective value. Also, we showed that the dynamic approach achieved higher approximation performance than the conventional weight vector distribution on problems with different Pareto front shapes without parameter tuning of the intermediate objective value. We focus on MOEA/D as a representative evolutionary algorithm for solving multi-objective optimization problems, I believe it is necessary to check whether it applies not only to MOEA/D but also to other decomposition base MOEAs including NSGA-III.

As future works, we will apply the proposed method to other decomposition-based algorithms such as NSGA-III and verify the effects on the search performance. We are designing an algorithm converting weight vector set based on multiple objective vectors.

References

1. Deb, K.: Multi-Objective Optimization using Evolutionary Algorithms. Wiley, New York (2001)
2. Coello, C.A.C., Veldhuizen, D.A.V., Lamont, G.B.: Evolutionary Algorithms for Solving Multi-Objective Problems. Kluwer Academic Publishers, Boston (2002)
3. Zhang, Q., Li, H.: MOEA/D: a multi-objective evolutionary algorithm based on decomposition. IEEE Trans. Evol. Comput. **11**(6), 712–731 (2007)

4. Jiang, S., Cai, Z., Zhang, J., Ong, Y.S.: Multiobjective optimization by decomposition with Pareto-adaptive weight vectors. In: Proceedings of 2011 Natural Computation (ICNC), pp. 1260–1264 (2011)

5. Deb, K., Jain, H.: An evolutionary many-objective optimization algorithm using reference-point based non-dominated sorting approach, Part II: handling constraints and extending to an adaptive approach. IEEE Trans. Evol. Comput. 18(4), 602–622 (2014)

6. Hamada, N., Nagata, Y., Kobayashi, S., Ono, I.: On the stopping criterion of adaptive weighted aggregation for multiobjective continuous optimization. Trans. Jpn. Soc. Evol. Comput. 4(1), 13–27 (2013)

7. Hamada, N., Nagata, Y., Kobayashi, S., Ono, I.: BS-AWA: a more scalable adaptive weighted aggregation for continuous multiobjective optimization. Trans. Jpn. Soc. Evol. Comput. 5(1), 1–15 (2014)

8. Sato, H.: Analysis of inverted PBI and comparison with other scalarizing functions in decomposition based MOEAs. J. Heuristics 21(6), 819–849 (2015)

9. Li, K., Zhang, Q., Kwong, S., Li, M., Wang, R.: Stable matching based selection in evolutionary multiobjective optimization. IEEE Trans. Evol. Comput. 18(6), 909–923 (2013)

10. Huband, S., Hingston, P., Barone, L., While, L.: A review of multi-objective test problems and a scalable test problem toolkit. IEEE Trans. Evol. Comput. 10(5), 477–506 (2006)

11. Zitzler, E.: Evolutionary Algorithms for Multiobjective Optimization: Methods and Applications. Ph.D. thesis, Swiss Federal Institute of Technology, Zurich (1999)

Classification of Permutation Distance Metrics for Fitness Landscape Analysis

Vincent A. Cicirello$^{(\boxtimes)}$ (ID)

Stockton University, Galloway, NJ 08205, USA
cicirelv@stockton.edu
https://www.cicirello.org/

Abstract. Commonly used computational and analytical tools for fitness landscape analysis of optimization problems require identifying a distance metric that characterizes the similarity of different solutions to the problem. For example, fitness distance correlation is Pearson correlation between solution fitness and distance to the nearest optimal solution. In this paper, we survey the available distance metrics for permutations, and use principal component analysis to classify the metrics. The result is aligned with existing classifications of permutation problem types produced through less formal means, including the A-permutation, R-permutation, and P-permutation types, and has also identified subtypes. The classification can assist in identifying appropriate metrics based on optimization problem feature for use in fitness landscape analysis. Implementations of all of the permutation metrics, and the code for our analysis, are available as open source.

Keywords: Fitness landscape analysis · Permutation distance · Permutation metric · Combinatorial optimization · Fitness distance correlation

1 Introduction

The concept of a fitness landscape originated in Mendelian genetics [24], and is now extensively used in the analysis of genetic algorithms and other forms of evolutionary computation. A fitness (or search) landscape [15] is the space of possible solutions to an optimization problem spatially arranged on a landscape with "similar" solutions neighboring each other, and where elevation corresponds to fitness (or solution quality). Peaks (for a maximization problem) and valleys (for a minimization problem) correspond to locally optimal solutions. The optimization problem is to find an optimal point on that landscape. Search landscape analysis is the term covering the theoretical and practical techniques for studying what characteristics of a problem make it hard, how different search operators affect fitness landscape topology, among others. In our research, we are especially interested in the fitness landscapes of permutation optimization problems, where solutions

© ICST Institute for Computer Sciences, Social Informatics and Telecommunications Engineering 2019
Published by Springer Nature Switzerland AG 2019. All Rights Reserved
A. Compagnoni et al. (Eds.): BICT 2019, LNICST 289, pp. 81–97, 2019.
https://doi.org/10.1007/978-3-030-24202-2_7

are represented by permutations of the elements of some set, and where we must maximize or minimize some function. For example, a solution to a traveling salesperson problem (TSP) is a permutation of the set of cities, and the objective is to find the permutation that corresponds to the minimal cost tour.

There is much work on fitness landscape analysis, including for permutation landscapes [3,6,8,19,21,22]. Fitness landscape analysis can use fitness distance correlation (FDC) [10], Pearson correlation between solution fitness and distance to the nearest optimal solution. The search landscape calculus [4] is another fitness landscape analysis tool that examines the local rate of change of fitness.

Fitness landscape analysis tools such as FDC and search landscape calculus require distance metrics. The features of a given structure, such as a permutation, that are important in determining similarity or distance is often problem dependent. For example, for a TSP, the permutation represents a set of edges between adjacent pairs of cities. Circularly rotate the permutation, and it still represents the same set of edges, and thus the same TSP solution. However, permutations can also represent one-to-one mappings between the elements of two sets. For example, in the largest common subgraph problem, one must find the largest subgraph (in number of edges) of graph G_1 that is isomorphic to a subgraph of graph G_2. Potential solutions to this problem can be represented by keeping the vertices of one of the graphs in a fixed order, and using a permutation of the vertices of the other graph to represent a mapping. In this case, if vertex i of G_2 is in position j of the permutation, it corresponds to mapping vertex i of G_2 to vertex j of G_1. In this example, it is the absolute positions of the elements in the permutation that are important to fitness. Campos et al. categorized permutation optimization problems into two types [1]: R-permutation problems, such as the TSP, where relative positions (i.e., adjacency implies edges) are important; and A-permutation problems, such as mapping problems, where absolute element positions have greatest effect on fitness. We previously added a third type, P-permutation, to this classification [4]. In a P-permutation problem general element precedences most directly impact solution fitness (e.g., element w occurs prior to elements x, y, and z, but not necessarily adjacent to any of them). Many scheduling problems fall into this class (e.g., a job x may be delayed if there are jobs with long process times anywhere prior to it in the schedule).

In this paper, we begin in Sect. 2 with a survey of the wide variety of permutation distance metrics that are described in the research literature. In Sects. 3 and 4 we then use principal component analysis (PCA) to formally identify groups of related permutation distance metrics from among those available. We will see that the first three principal components correspond to the three problem classes previously defined; and that our approach additionally identifies subtypes. A classification of permutation distance metrics that aligns with the existing classification of permutation problems is a desirable property of our results. For example, if one requires a permutation metric relevant for analyzing the fitness landscape of a problem known to be in a particular problem class, then the distance classification can directly lead to the most relevant metrics. Next, in Sect. 5, we provide a set of fitness landscapes that correspond to the identified classes of permutation distance metric. For each of these landscapes and for each metric, we compute FDC as an example application of the

Table 1. Summary of distance measure classes.

Permutation distance	Runtime	Metric?
Edit distance	$O(n^2)$	Yes
Exact match distance	$O(n)$	Yes
Interchange distance	$O(n)$	Yes
Acyclic edge distance	$O(n)$	Pseudo
Cyclic edge distance	$O(n)$	Pseudo
R-type distance	$O(n)$	Yes
Cyclic r-type distance	$O(n)$	Pseudo
Reversal edit distance	Init: $O(n!n^3)$ Compute: $O(n^2)$	Yes
Kendall tau distance	$O(n \lg n)$	Yes
Reinsertion distance	$O(n \lg n)$	Yes
Deviation distance	$O(n)$	Yes
Normalized deviation distance	$O(n)$	Yes
Squared deviation distance	$O(n)$	Yes
Lee distance	$O(n)$	Yes

classification scheme. We implement the PCA, as well as our FDC examples, in Java, using an open source Java library of permutation distance metrics [5]. We have added the source code for our analysis to the repository to enable easily replicating our results. The source repository is found at https://github.com/cicirello/JavaPermutationTools, and additional documentation for the library itself at https://jpt.cicirello.org/. We wrap up with a discussion of the classification in Sect. 6.

2 Permutation Distance

Table 1 summarizes the permutation distance metrics used in our analysis, including runtime, and indicating which are metrics. The n in runtimes and in equations is permutation length. Wherever we specify a distance mathematically, $p(i)$ refers to the element in position i of permutation p; and we use 1-based indexing in the equations (index of first position of a permutation is 1). If we need to refer to two different permutations, we use subscripts. Thus, $p_1(i)$ refers to the element in position i of permutation p_1.

Edit Distance: The edit distance between two structures is the minimum cost of the "edit operations" required to transform one structure into the other. Levenshtein distance is a string edit distance [13], where the edit operations are inserting a new character, removing an existing character, or changing a character to a different one. Levenshtein was concerned with binary strings (i.e., of ones and zeros). Wagner and Fischer extended this to non-binary strings, introduced the ability to apply different costs to the three types of edit operations,

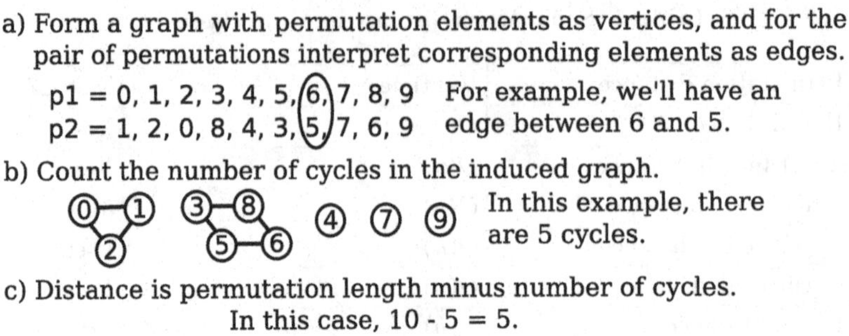

a) Form a graph with permutation elements as vertices, and for the pair of permutations interpret corresponding elements as edges.

p1 = 0, 1, 2, 3, 4, 5, 6, 7, 8, 9   For example, we'll have an
p2 = 1, 2, 0, 8, 4, 3, 5, 7, 6, 9   edge between 6 and 5.

b) Count the number of cycles in the induced graph.

In this example, there are 5 cycles.

c) Distance is permutation length minus number of cycles.
In this case, 10 - 5 = 5.

Fig. 1. Computing interchange distance via cycle counting.

and provided a dynamic programming algorithm for computing it [23]. Sörensen suggested treating a permutation as a string, and applying string edit distance to permutations [21]. All edit distances are metrics. Our edit distance implementation is of Wagner and Fischer's dynamic programming algorithm, including parameters for the costs of the edit operations. Runtime is $O(n^2)$.

Exact Match Distance: Ronald extended Hamming distance to non-binary strings, producing a permutation distance he called exact match distance [18], which is the number of positions with different elements. It is an edit distance where the only edit operation is element changes. It is widely used [3,6,20,21], satisfies the metric properties [18], and has runtime $O(n)$. We define it as:

$$\delta(p_1, p_2) = \sum_{i=1}^{n} \begin{cases} 1 \text{ if } p_1(i) \neq p_2(i) \\ 0 \text{ otherwise.} \end{cases} \tag{1}$$

Interchange Distance: Interchange distance is an edit distance with one edit operation, element interchanges (or swaps). It is the minimum number of swaps needed to transform p_1 into p_2; and is computed efficiently ($O(n)$ time) by counting the number of cycles between the permutations [6]. A permutation cycle of length k is transformed into k fixed points with $k - 1$ swaps (a fixed point is a cycle of length 1). Figure 1 illustrates computing interchange distance by cycle counting. Let CycleCount(p_1, p_2) be the number of permutation cycles. Thus, we formalize interchange distance as:

$$\delta(p_1, p_2) = n - \text{CycleCount}(p_1, p_2). \tag{2}$$

Cyclic Edge Distance and Acyclic Edge Distance: Ronald defines measures useful when permutations represent sets of edges: cyclic edge distance and acyclic edge distance [16,17]. Both assume that the element adjacency within a permutation correspond to undirected edges. Cyclic edge distance considers the permutation to be a cycle, where the first and last elements are adjacent; whereas acyclic edge distance does not. Cyclic edge distance interprets the permutation, $[0, 1, 2, 3, 4]$, as the set of undirected edges, $\{(0, 1), (1, 2), (2, 3), (3, 4), (4, 0)\}$, while acyclic edge distance excludes $(4, 0)$ from this set. Both are invariant under a complete reversal (e.g., $[0, 1, 2, 3, 4]$ is equivalent to $[4, 3, 2, 1, 0]$). The cyclic form is also invariant under rotations. In both forms, distance is the number of edges that are not in common (i.e., the number of edges in p_1 that are not found in p_2) and is computed in $O(n)$ time. Both are pseudo-metrics [17] (due to reversal invariance, and rotational invariance for the cyclic form). We formalize cyclic and acyclic edge distances, respectively, as follows:

$$\delta(p_1, p_2) = \sum_{i=1}^{n} \begin{cases} 0 & \text{if } \exists j \exists x \exists y, \; j = (i \bmod n) + 1 \wedge y = (x \bmod n) + 1 \wedge \\ & \quad [(p_1(i) = p_2(x) \wedge p_1(j) = p_2(y)) \\ & \quad \vee (p_1(i) = p_2(y) \wedge p_1(j) = p_2(x))] \\ 1 & \text{otherwise.} \end{cases} \tag{3}$$

$$\delta(p_1, p_2) = \sum_{i=1}^{n-1} \begin{cases} 0 & \text{if } \exists x, \; (p_1(i) = p_2(x) \; \wedge p_1(i+1) = p_2(x+1)) \vee \\ & \quad (p_1(i) = p_2(x+1) \; \wedge p_1(i+1) = p_2(x)) \\ 1 & \text{otherwise.} \end{cases} \tag{4}$$

R-Type Distance and Cyclic R-Type Distance: The r-type distance ("r" for relative) [1] is a directed edge version of acyclic edge distance. Cyclic r-type distance [4] is a cyclic counterpart to r-type distance, which includes an edge between the end points. Though r-type distance satisfies the metric properties, cyclic r-type is a pseudo-metric due to rotational invariance. Both are computed in $O(n)$ time, and defined respectively as:

$$\delta(p_1, p_2) = \sum_{i=1}^{n-1} \begin{cases} 0 & \text{if } \exists x, \; p_1(i) = p_2(x) \; \wedge p_1(i+1) = p_2(x+1) \\ 1 & \text{otherwise.} \end{cases} \tag{5}$$

$$\delta(p_1, p_2) = \sum_{i=1}^{n} \begin{cases} 0 & \text{if } \exists j \exists x \exists y, \; j = (i \bmod n) + 1 \wedge y = (x \bmod n) + 1 \wedge \\ & \quad p_1(i) = p_2(x) \wedge p_1(j) = p_2(y) \\ 1 & \text{otherwise.} \end{cases} \tag{6}$$

Reversal Edit Distance: Reversal edit distance is the minimum number of reversals needed to transform p_1 into p_2. Computing reversal edit distance is NP-Hard [2]; and Schiavinotto and Stützle argue that the best available approximations are insufficient for search landscape analysis [19].

Our implementation of reversal edit distance uses breadth-first enumeration to initialize a lookup table mapping each of the $n!$ permutations to its reversal edit distance from a reference permutation. Later, computing the distance between a given pair of permutations becomes a table lookup. We originally implemented this for a context where we required computing distance from all permutations of a specific relatively short length ($n = 10$) to one specific permutation [4]. In that context, initialization cost is $O(n!n^3)$ (i.e., breadth-first enumeration iterates over $O(n!)$ permutations, each of which has $O(n^2)$ neighbors (i.e., possible sub-permutation reversals), and the cost to execute a reversal is $O(n)$. Therefore, applications with the need to compute $O(n!)$ distances all from the same reference permutation can do so with an amortized initialization cost of $O(n^3)$ per distance calculation. The table lookup has cost $O(n^2)$ (cost to compute mixed radix representation of the permutation).

Kendall Tau Distance: Kendall tau distance, a metric, is a slight variation of Kendall's rank correlation coefficient [11]:

$$\delta(p_1, p_2) = \sum_{i=1}^{n-1} \sum_{j=(i+1)}^{n} \begin{cases} 0 & \text{if } \exists x \exists y, \ p_1(i) = p_2(x) \ \wedge p_1(j) = p_2(y) \ \wedge \ x < y \\ 1 & \text{otherwise.} \end{cases} \quad (7)$$

Some divide this sum by $n(n-1)/2$, but most use it in the form of Eq. 7 (e.g., [7, 14]) where it corresponds to the minimum number of adjacent swaps needed to transform permutation p_1 into p_2. Thus, it is an adjacent swap edit distance. For this reason, it is sometimes called bubble sort distance, since it corresponds to the number of adjacent swaps executed by bubble sort. The runtime of our implementation of Kendall tau distance is $O(n \lg n)$ using a modified version of mergesort to count inversions.

Reinsertion Distance: Reinsertion distance is an edit distance with a single atomic edit operation, removal/reinsertion, which removes an element and reinserts it elsewhere in the permutation; and thus is the minimum number of removal/reinsertions needed to transform p_1 into p_2. Relying on the observation that the elements that must be removed and reinserted are exactly the elements that do not lie on the longest common subsequence [4], it can be computed efficiently in $O(n \lg n)$ time (e.g., using Hunt and Szymanski's algorithm for longest common subsequence [9]). Thus, we implement reinsertion distance as:

$$\delta(p_1, p_2) = n - \#(\text{MaxCommonSubsequence}(p_1, p_2)). \quad (8)$$

Deviation Distance and Normalized Deviation Distance: Deviation distance is the sum of the positional deviations of the permutation elements, and is a metric [18]. The positional deviation of an element is the absolute value of the difference of its index in p_1 from its index in p_2. Ronald [18] originally divided this sum by $n-1$ to bound an element's contribution to total distance in the interval $[0, 1]$. Many use this form (e.g., [21]) including in our own prior work [3,6]. Others (e.g., [1,20]), including our own prior work [4], do not divide by $(n-1)$. Runtime of our implementation is $O(n)$. The two forms are:

$$\delta(p_1, p_2) = \frac{1}{n-1} \sum_{e \in p_1} |i - j|, \text{ where } p_1(i) = p_2(j) = e. \quad (9)$$

$$\delta(p_1, p_2) = \sum_{e \in p_1} |i - j|, \text{ where } p_1(i) = p_2(j) = e. \quad (10)$$

Squared Deviation Distance: Sevaux and Sörensen suggested squared deviation distance, which is based on Spearman's rank correlation coefficient [20].

It is the sum of the squares of the positional deviations of the permutation elements. Sevaux and Sörensen falsely state that squared deviation distance as well as deviation distance require quadratic time [20], however our implementations of these are $O(n)$ time, with two linear passes, the first to generate the inverse of one permutation, which is then used in the second pass as a lookup table (i.e., to find element indices).

$$\delta(p_1, p_2) = \sum_{e \in p_1} (i - j)^2, \text{where } p_1(i) = p_2(j) = e. \tag{11}$$

Table 2. Lower triangle of correlation matrix (columns in same order as rows).

Exact match	1.000									
Interchange	0.766	1.000								
Acyclic edge	0.019	0.070	1.000							
Cyclic edge	−0.000	0.056	0.899	1.000						
Rtype	0.024	0.009	0.628	0.564	1.000					
Cyclic rtype	−0.000	−0.010	0.557	0.619	0.911	1.000				
Kendall tau	0.328	0.241	−0.000	0.000	0.085	0.075	1.000			
Reinsertion	0.301	0.182	0.102	0.100	0.422	0.392	0.704	1.000		
Deviation (dev)	0.515	0.395	0.008	−0.000	0.020	−0.000	0.931	0.650	1.000	
Squared dev	0.333	0.255	−0.000	−0.000	0.017	−0.000	0.984	0.623	0.947	1.000
Lee	0.556	0.426	0.019	0.000	0.014	−0.000	0.447	0.452	0.703	0.455	1.000

Lee Distance: We include in our analysis an adaptation for permutations of Lee distance [12] for strings, which originated in coding theory. Lee distance sums the positional deviations of the elements, however, it uses the minimum of the deviations to the left and right treating the permutation as a cyclic structure. It is a metric, and is computed in $O(n)$ time. Define it as:

$$\delta(p_1, p_2) = \sum_{e \in p_1} \min(|i - j|, n - |i - j|), \text{where } p_1(i) = p_2(j) = e. \tag{12}$$

3   Classification of Permutation Distance Measures

We use principal component analysis to identify groups of related permutation distance metric. We use all of the distance measures from Sect. 2 except edit distance, normalized deviation distance, and reversal edit distance. We exclude edit distance because its parameters define a continuum of distance metrics. We exclude normalized deviation distance because it is simply deviation distance scaled, and thus any observations made of deviation distance apply to both. We exclude reversal edit distance due to cost of computing it, however, we later discuss where it fits in our classification. We begin by generating a dataset by iterating over all permutations of length $n = 10$ and computing distances to a single reference permutation. We then compute the correlation matrix, found in Table 2.

Using Jacobi iteration, we compute the eigenvalues and eigenvectors of the correlation matrix. Table 3 lists the eigenvalues of the principal components (PC). The first three PCs have eigenvalues greater than 1.0; and the first five PCs combine for greater than 90% of the sum. Table 4 provides the eigenvectors associated with the first five PCs. Table 5 lists the correlation between the original distance metrics and each of the first five PCs. The first three PCs (all with eigenvalues greater than 1) correspond to the three types of permutation optimization problem discussed earlier in Sect. 1.

Table 3. Eigenvalues of the principal components.

PC	Eigenvalue	Proportion	Cumulative
1	4.3644	0.3968	0.3968
2	3.1148	0.2832	0.6799
3	1.4740	0.1340	0.8139
4	0.8367	0.0761	0.8900
5	0.5465	0.0497	0.9397
6	0.2492	0.0227	0.9623
7	0.2120	0.0193	0.9816
8	0.1575	0.0143	0.9959
9	0.0315	0.0029	0.9988
10	0.0107	0.0010	0.9998
11	0.0026	0.0002	1.0000

PC1 (P-Permutation): PC1 correlates extremely strongly (0.94) to deviation distance, very strongly to Kendall tau distance and squared deviation distance, and reasonably strongly to reinsertion distance and Lee distance (Table 5). The Kendall tau and reinsertion distances, by their very definitions, focus on permutation similarity in terms of pairwise element precedences. Although the variations of deviation distance do not explicitly consider this, they capture that essence in that an element that is displaced a greater number of positions is likely involved in a greater number of precedence inversions (i.e., where a is prior to b in one permutation, and somewhere after b in the other). So these five permutation metrics are P-permutation distances, measuring permutation distance in terms of precedence related features.

Table 4. Eigenvectors of the first five principal components.

Distance	PC1	PC2	PC3	PC4	PC5
Exact match distance	0.2984	0.0958	0.5419	−0.1573	0.1423
Interchange distance	0.2487	0.0695	0.6058	−0.0586	0.3936
Acyclic edge distance	0.0854	−0.4751	0.1354	0.4611	−0.0635
Cyclic edge distance	0.0805	−0.4768	0.1194	0.4674	−0.0455
R-type distance	0.1271	−0.4873	−0.0576	−0.3803	0.0517
Cyclic r-type distance	0.1153	−0.4874	−0.0666	−0.3793	0.0510
Kendall tau distance	0.4216	0.0928	−0.3110	0.1400	0.2292
Reinsertion distance	0.3721	−0.0848	−0.2529	−0.3795	−0.0509
Deviation distance	0.4516	0.1321	−0.1089	0.1630	−0.0651
Squared deviation distance	0.4140	0.1189	−0.2828	0.2444	0.2218
Lee distance	0.3381	0.1027	0.2157	−0.0476	−0.8396

PC2 (R-Permutation): PC2 correlates very strongly with both forms of edge distance, and both forms of R-type distance ($|r| > 0.83$ in all four cases). These distances all focus on adjacency (i.e., edges) of permutation elements.

Table 5. Correlation between distance metrics and first five principal components.

Distance	PC1	PC2	PC3	PC4	PC5
Exact match distance	0.6234	0.1691	0.6579	−0.1439	0.1052
Interchange distance	0.5196	0.1227	0.7355	−0.0536	0.2910
Acyclic edge distance	0.1784	−0.8385	0.1644	0.4218	−0.0470
Cyclic edge distance	0.1682	−0.8415	0.1450	0.4276	−0.0337
R-type distance	0.2654	−0.8600	−0.0699	−0.3479	0.0382
Cyclic r-type distance	0.2410	−0.8602	−0.0808	−0.3469	0.0377
Kendall tau distance	0.8808	0.1638	−0.3775	0.1281	0.1695
Reinsertion distance	0.7774	−0.1497	−0.3070	−0.3472	−0.0377
Deviation distance	0.9435	0.2332	−0.1322	0.1491	−0.0481
Squared deviation distance	0.8649	0.2099	−0.3434	0.2236	0.1640
Lee distance	0.7063	0.1812	0.2619	−0.0436	−0.6207

PC3 (A-Permutation): PC3 strongly correlates to exact match distance and interchange distance ($r = 0.6579$ and $r = 0.7355$, respectively). Both of these distance metrics focus on absolute positions of permutation elements.

The fourth and fifth PCs identify subtypes. Their eigenvalues are less than 1, and account for relatively small portions of the eigenvalue sum (approximately 7.6% and 5%), but is interesting to interpret their structure none-the-less.

Table 6. Permutation distance metric classification.

Type	Subtype	Distance
P-permutation	Acyclic subtype	Kendall tau distance, reinsertion distance, deviation distance, squared deviation distance
	Cyclic subtype	Lee distance
R-permutation	Undirected subtype	Acyclic edge distance, cyclic edge distance, reversal edit distance
	Directed subtype	R-type distance, cyclic r-type distance
A-permutation		Exact match distance, interchange distance

PC4 (R-Permutation, Undirected Subtype): PC4's strongest correlations are to the two variations of edge distance, which consider permutations to represent sets of undirected edges.

PC5 (P-Permutation, Cyclic Subtype): PC5 has moderately strong correlation ($r = -0.6207$) to Lee distance, and only weak correlation to the other distances. Lee distance also had strong correlation with PC1 (P-permutation), however, Lee distance is different than the other metrics based on deviations in that the positional deviation is computed as if the end points are linked. So in some sense, we might consider this a cyclic subtype of P-permutation.

Our classification of the distance metrics is found in Table 6. It includes three primary types: P-permutation, R-permutation, and A-permutation; and subdivides two of the types into subtypes. Although we excluded reversal edit distance in the analysis, we include it among the undirected R-permutation metrics as a reversal operation essentially replaces two undirected edges.

Table 7. Lower triangle of correlation matrix (permutation length 50).

Exact match	1.000										
Interchange	0.578	1.000									
Acyclic edge	0.001	0.009	1.000								
Cyclic edge	0.000	0.009	0.980	1.000							
Rtype	0.001	0.000	0.693	0.679	1.000						
Cyclic rtype	−0.000	−0.000	0.679	0.693	0.980	1.000					
Kendall tau	0.142	0.082	−0.000	−0.000	0.008	0.007	1.000				
Reinsertion	0.140	0.074	0.060	0.059	0.176	0.172	0.532	1.000			
Deviation (dev)	0.226	0.132	−0.000	−0.000	−0.001	−0.001	0.944	0.555	1.000		
Squared dev	0.143	0.084	−0.000	−0.000	−0.000	−0.001	0.995	0.501	0.949	1.000	
Lee	0.248	0.144	0.000	−0.000	−0.001	−0.001	0.431	0.439	0.685	0.433	1.000

Table 8. Eigenvalues of the principal components (permutation length 50).

PC	Eigenvalue	Proportion	Cumulative
1	3.7755	0.3432	0.3432
2	3.3513	0.3047	0.6479
3	1.5162	0.1378	0.7857
4	0.7515	0.0683	0.8541
5	0.6604	0.0600	0.9141
6	0.4849	0.0441	0.9582
7	0.4111	0.0374	0.9955
8	0.0336	0.0031	0.9986
9	0.0069	0.0006	0.9992
10	0.0059	0.0005	0.9998
11	0.0027	0.0002	1.0000

4 On the Relevance to Longer Permutations

In the PCA conducted in Sect. 3 to generate our classification scheme, we computed the correlations for the correlation matrix using permutations of length $n = 10$ and iterated over all permutations of that length. To explore whether permutation length has an effect on the classes identified, in this section we repeat the PCA using permutations of length $n = 50$. This length is too long to compute the correlations using all permutations, so instead we randomly sample the space of permutations. We use 3628800 randomly sampled permutations of length 50 (the size of the space of permutations of length 10 so our correlations are computed using the same number of data points as in Sect. 3. Table 7 shows the correlation matrix. Table 8 provides the eigenvalues, and Table 9 shows the eigenvectors of the first five principal components. Table 10 shows the correlations between the original distance metrics and each of the first five principal components.

From Table 10, we again see that PC1 correlates extremely strongly to deviation distance, Kendall tau distance, and squared deviation distance ($|r| > 0.9$ in those cases), and also correlates strongly to reinsertion distance and Lee distance. PC1, as before, corresponds to the P-permutation metrics. Likewise, PC2 (as before) correlates very strongly ($|r| > 0.89$) to both forms of edge distance and both forms of R-type distance; and thus corresponds to the R-permutation metrics. PC3 correlates very strongly to both exact match distance ($r = -0.8265$) and interchange distance ($r = -0.8525$). This likewise is consistent with the results for shorter length permutations, and corresponds to the the A-permutation metrics. PC5 again correlates moderately strongly to Lee distance ($r = -0.5258$) and only weakly to the others.

Table 9. Eigenvectors of the first five principal components (permutation length 50).

Distance	PC1	PC2	PC3	PC4	PC5
Exact match distance	−0.1601	−0.0393	−0.6712	0.0106	0.0194
Interchange distance	−0.1125	−0.0254	−0.6923	0.1506	0.1593
Acyclic edge distance	−0.0954	0.4879	−0.0109	0.2416	−0.3347
Cyclic edge distance	−0.0951	0.4879	−0.0103	0.2424	−0.3348
R-type distance	−0.1089	0.4878	0.0055	−0.1944	0.3089
Cyclic r-type distance	−0.1084	0.4878	0.0060	−0.1924	0.3079
Kendall tau distance	−0.4675	−0.1104	0.1629	0.3089	0.1655
Reinsertion distance	−0.3550	0.0016	0.0632	−0.5557	0.2897
Deviation distance	−0.4918	−0.1188	0.0827	0.0987	−0.0987
Squared deviation distance	−0.4648	−0.1129	0.1608	0.3356	0.1418
Lee distance	−0.3428	−0.0814	−0.0815	−0.5086	−0.6471

Table 10. Correlation between distance metrics and first five principal components.

Distance	PC1	PC2	PC3	PC4	PC5
Exact match distance	−0.3111	−0.0720	−0.8265	0.0092	0.0157
Interchange distance	−0.2185	−0.0464	−0.8525	0.1306	0.1294
Acyclic edge distance	−0.1853	0.8932	−0.0134	0.2094	−0.2720
Cyclic edge distance	−0.1849	0.8932	−0.0127	0.2102	−0.2720
R-type distance	−0.2116	0.8931	0.0067	−0.1685	0.2510
Cyclic r-type distance	−0.2106	0.8931	0.0074	−0.1668	0.2502
Kendall tau distance	−0.9083	−0.2021	0.2006	0.2678	0.1345
Reinsertion distance	−0.6898	0.0030	0.0778	−0.4817	0.2355
Deviation distance	−0.9555	−0.2175	0.1019	0.0855	−0.0802
Squared deviation distance	−0.9032	−0.2067	0.1980	0.2910	0.1152
Lee distance	−0.6661	−0.1490	−0.1003	−0.4409	−0.5258

PC4 is the only inconsistency when conducting the PCA using longer permutations (length 50) and randomly sampling the space of permutations as compared to shorter permutations (length 10) and computing the correlations using all permutations of that length. Before, with shorter permutations, PC4 identified the two forms of edge distance, which we referred to as R-permutation undirected subtype. With longer permutations that are randomly sampled, PC4 has identified reinsertion distance, and to a lesser extent Lee distance. This suggests that as permutation length is increased that there may be a relationship between reinsertion distance and Lee distance; or at the very least that reinsertion distance captures a rather different essence of permutation variability than does the other P-permutation metrics.

We have chosen to stick with the classification identified earlier in Table 6, since four of the five PCAs directly correspond to that earlier analysis, and since the specifics of the distinct nature of reinsertion distance are not entirely clear.

5   Example Fitness Landscapes

In this section, we examine five search landscapes as examples.

R-Permutation Landscape, Undirected Subtype (L_1): The first search landscape is for a simple instance of the TSP with a known optimal solution. Specifically, it consists of 20 cities arranged on a circle with radius 1.0, with equidistant separation between each consecutive pair of cities. The cost of an edge is Euclidean distance. The optimal solution is to either follow the cities around the circle clockwise or counterclockwise returning to the starting city to complete the tour. In the space of permutations, there are 40 optimal solutions: 20 cities at which the permutation can begin, and two possible travel directions (clockwise and counterclockwise).

In Table 11 we provide FDC computed using 100000 randomly sampled permutations. FDC is Pearson correlation between the fitness of a solution to the problem and the distance to the nearest optimal solution. In this case, it is the correlation between the cost of the tour of cities that the permutation represents, and the distance to the nearest of the 40 optimal permutations. We have used boldface font in the table to make it easy to see where we found the highest FDC. Specifically, the highest FDC was seen for the two forms of edge distance, and it was also reasonably high for the two forms of R-type (recall that the R-type distance uses directed edges, while edge distance uses undirected edges). Cyclic edge distance had slightly higher FDC over acyclic edge distance, which makes sense since a solution to a TSP is a cycle of the cities so that the first and last elements of the permutation represents an edge.

R-Permutation Landscape, Directed Subtype (L_2): The second landscape is for a simple asymmetric TSP (ATSP) instance. We again have 20 cities arranged on a circle of radius 1.0, with equidistant separation around the circle. Let city c_0 be the city at "three o'clock" on the circle, and let city c_i be the next city after c_{i-1} in counterclockwise order around the circle. The cost of the edge from city c_i to city c_j is Euclidean distance if $i < j$, and is otherwise a constant distance 2.0 if $i > j$. There is one optimal tour for this instance of the ATSP, which is to visit the cities in counterclockwise order. In the space of permutations, there are 20 optimal solutions that correspond to this tour: 20 starting cities.

For this landscape, we find that the two forms of R-type distance offer the highest FDC (see Table 11) and that FDC is otherwise low for the other permutation distance measures. The cyclic form has slightly higher FDC than the acyclic form, which is consistent with the cyclic nature of ATSP solutions.

A-Permutation Landscape (L_3): Our example A-permutation search landscape is a variation of the "Permutation in a Haystack" problem [4]. An instance

Table 11. Fitness-distance correlation for five example landscapes and each measure of distance.

Distance	L_1	L_2	L_3	L_4	L_5
Exact match distance	0.1548	0.1881	**0.6917**	0.2974	0.4806
Interchange distance	0.1192	0.0886	**0.5296**	0.2204	0.3665
Acyclic edge distance	**0.6052**	0.3474	0.0118	0.0020	0.0186
Cyclic edge distance	**0.6204**	0.3822	−0.0002	0.0006	0.0026
R-type distance	**0.5442**	**0.6333**	0.0148	0.0790	0.0136
Cyclic r-type distance	**0.5562**	**0.6595**	−0.0016	0.0684	0.0005
Kendall tau distance	0.3423	0.2408	0.2245	**0.9022**	0.3862
Reinsertion distance	0.3382	0.5349	0.2080	**0.6364**	0.3887
Deviation distance	0.3898	0.1875	0.3544	**0.8410**	0.6072
Squared deviation distance	0.3150	0.1555	0.2282	**0.8876**	0.3935
Lee distance	0.4640	0.2316	0.3836	0.4063	**0.8619**

of the "Permutation in a Haystack" problem is defined by specifying the optimal permutation p, and then defining the optimization objective as minimizing the distance to p for some specific choice of permutation distance metric. It is the permutation analog of the "OneMax" fitness landscape often used in testing genetic algorithms with the bitstring representation. The "Permutation in a Haystack" problem enables easily defining permutation optimization landscapes for testing and experimentation purposes that possess the topology that you wish to study along with a known optimal solution.

For landscape L_3, we modify the "Permutation in a Haystack" slightly. Specifically, rather than using a distance function (as is) for the optimization objective, we instead use a noisy distance function. After choosing p, the fitness of a permutation q in landscape L_3 is equal to $\alpha_q * \delta(p, q)$, where $\delta(p, q)$ is the exact match distance between q and the optimal solution p, and the α_q values are generated uniformly at random from the interval $[1, 1.5)$. We use a slightly smaller permutation length of 10 for L_3 than we did for the first two.

In Table 11, you will see that the two permutation metrics that were earlier identified as A-permutation by our PCA both have high FDC to landscape L_3; and FDC is low for all other permutation metrics.

P-Permutation Landscape, Acyclic Subtype (L_4): We use the same variation of the "Permutation in a Haystack" problem as in L_3 to obtain a P-permutation landscape with a known optimal solution. Specifically, the fitness of permutation q is equal to $\alpha_q * \delta(p, q)$, but this time $\delta(p, q)$ is the Kendall tau distance between q and the optimal solution p.

In Table 11, we find that three of the four permutation metrics that we classified as the acyclic subtype of the P-permutation class have very high FDC for landscape L_4 (namely, the Kendall tau, deviation, and squared deviation distance metrics). The fourth, reinsertion distance, also has a reasonably high FDC for this landscape; while all other distance measures have low FDC.

P-Permutation Landscape, Cyclic Subtype (L_5): The last of our example permutation fitness landscapes uses Lee distance in the variation of the "Permutation in a Haystack" problem that we used in landscapes L_3 and L_4. You can see in Table 11 that Lee distance provides the highest FDC for this landscape.

The distance metrics that lead to highest FDC for each of these five example fitness landscapes correspond to the metrics from each of the five classifications from Table 6, the three primary classes along with the subtypes. If we additionally had a mapping of the available search operators (e.g., mutation and crossover operators for a genetic algorithm, neighborhood operators for simulated annealing and other local search) to the classification scheme, then it would assist in selecting relevant operators for the optimization problem at hand.

6   Discussion and Conclusions

In this paper, we used PCA to produce a classification of distance metrics for permutations. The analysis used all of the most common permutation distance metrics, providing open source implementations in the Java language. The code for our PCA is likewise available in that same repository to enable others to easily replicate our results.

The classification can help in the selection of a distance metric for use in fitness distance correlation or other fitness landscape analysis techniques. For example, if you are analyzing a search landscape for a problem where permutations represent sets of edges (e.g., TSP and similar problems), then the classification would suggest use of either one of the forms of edge distance or r-type distance, depending upon whether the edges are undirected (like the TSP) or directed (like the asymmetric TSP). Or, if you are faced with a P-permutation problem, one where general pairwise element precedences have the greatest impact on fitness (e.g., many scheduling problems), then you would instead choose a P-permutation metric such as Kendall tau distance, reinsertion distance, or one of the variations of deviation distance. Additionally, in this case, you may then factor in the runtimes of the alternative metrics in your choice. For example, Kendall tau distance and squared deviation distance correlate very strongly ($r = 0.984$, Table 2). However, Kendall tau distance is computed in time $O(n \lg n)$, while squared deviation distance is computed in $O(n)$ time. Even if Kendall tau is the best fit for your specific analysis problem, squared deviation may be sufficient due to its strong correlation while saving computational cost.

Another potential use in fitness landscape analysis is in identifying search operators most relevant to a problem. For example, for some sample instances of the search problem, you might start by computing fitness distance correlation using one (or more) distance metric(s) from each class. Essentially this step would map your problem into the same classification. Identifying the class of the problem can then assist in identifying relevant search operators. For simulated annealing, and as the mutation operator for a genetic algorithm, an insertion operator has been shown quite effective in general for P-permutation problems [4]. Insertion removes a random element of the permutation, and reinserts it at a different random point. While for an R-permutation problem, you might instead use a reversal operator (reverses a randomly chosen sub-permutation) or a block move (removes a random sub-permutation and reinserts it at a randomly chosen position). Reversals essentially replace two undirected edges, and block moves replace three edges. This approach can also be useful for identifying relevant crossover operators from among the many permutation crossover operators (cycle crossover, order crossover, etc) that are available, some of which better maintain edges while others better maintain absolute positions.

References

1. Campos, V., Laguna, M., Martí, R.: Context-independent Scatter and Tabu search for permutation problems. INFORMS Comput. **17**(1), 111–122 (2005)
2. Caprara, A.: Sorting by reversals is difficult. In: Proceedings of the First International Conference on Computational Molecular Biology, pp. 75–83. ACM (1997)
3. Cicirello, V.A.: On the effects of window-limits on the distance profiles of permutation neighborhood operators. In: Proceedings of the International Conference on Bioinspired Information and Communications Technologies, pp. 28–35, December 2014. https://doi.org/10.4108/icst.bict.2014.257872
4. Cicirello, V.A.: The permutation in a haystack problem and the calculus of search landscapes. IEEE Trans. Evol. Comput. **20**(3), 434–446 (2016). https://doi.org/10.1109/TEVC.2015.2477284
5. Cicirello, V.A.: JavaPermutationTools: a Java library of permutation distance metrics. J. Open Source Softw. **3**(31), 950 (2018). https://doi.org/10.21105/joss.00950
6. Cicirello, V.A., Cernera, R.: Profiling the distance characteristics of mutation operators for permutation-based genetic algorithms. In: Proceedings of the 26th International Conference of the Florida Artificial Intelligence Research Society, pp. 46–51. AAAI Press, May 2013
7. Fagin, R., Kumar, R., Sivakumar, D.: Comparing top k lists. SIAM J. Discret. Math. **17**(1), 134–160 (2003)
8. Hernando, L., Mendiburu, A., Lozano, J.A.: A tunable generator of instances of permutation-based combinatorial optimization problems. IEEE Trans. Evol. Comput. **20**(2), 165–179 (2016)
9. Hunt, J.W., Szymanski, T.G.: A fast algorithm for computing longest common subsequences. Commun. ACM **20**(5), 350–353 (1977)
10. Jones, T., Forrest, S.: Fitness distance correlation as a measure of problem difficulty for genetic algorithms. In: Proceedings of the 6th International Conference on Genetic Algorithms, pp. 184–192. Morgan Kaufmann (1995)
11. Kendall, M.G.: A new measure of rank correlation. Biometrika **30**(1/2), 81–93 (1938)
12. Lee, C.: Some properties of nonbinary error-correcting codes. IRE Trans. Inf. Theory **4**(2), 77–82 (1958)
13. Levenshtein, V.I.: Binary codes capable of correcting deletions, insertions and reversals. Sov. Phys. Dokl. **10**(8), 707–710 (1966)
14. Meilă, M., Bao, L.: An exponential model for infinite rankings. J. Mach. Learn. Res. **11**, 3481–3518 (2010)
15. Mitchell, M.: An Introduction to Genetic Algorithms. MIT Press, Cambridge (1998)

16. Ronald, S.: Finding multiple solutions with an evolutionary algorithm. In: IEEE Congress on Evolutionary Computation, pp. 641–646. IEEE Press (1995)
17. Ronald, S.: Distance functions for order-based encodings. In: Proceedings of the IEEE Congress on Evolutionary Computation, pp. 49–54. IEEE Press (1997)
18. Ronald, S.: More distance functions for order-based encodings. In: Proceedings of the IEEE Congress on Evolutionary Computation, pp. 558–563. IEEE Press (1998)
19. Schiavinotto, T., Stützle, T.: A review of metrics on permutations for search landscape analysis. Comput. Oper. Res. **34**(10), 3143–3153 (2007)
20. Sevaux, M., Sörensen, K.: Permutation distance measures for memetic algorithms with population management. In: Proceedings of the Metaheuristics International Conference (MIC 2005), pp. 832–838, August 2005
21. Sörensen, K.: Distance measures based on the edit distance for permutation-type representations. J. Heuristics **13**(1), 35–47 (2007)
22. Tayarani-N, M.H., Prugel-Bennett, A.: On the landscape of combinatorial optimization problems. IEEE Trans. Evol. Comput. **18**(3), 420–434 (2014). https://doi.org/10.1109/TEVC.2013.2281502
23. Wagner, R.A., Fischer, M.J.: The string-to-string correction problem. J. ACM **21**(1), 168–173 (1974)
24. Wright, S.: Evolution in Mendelian populations. Genetics **16**(2), 97–159 (1931)

Medical Diagnostics Based on Encrypted Medical Data

Alexey Gribov[1], Kelsey Horan[1,2(✉)], Jonathan Gryak[2], Kayvan Najarian[2], Vladimir Shpilrain[1], Reza Soroushmehr[2], and Delaram Kahrobaei[1,3]

[1] The Graduate Center, The City University of New York,
365 Fifth Ave, New York, NY 10016, USA
[2] Michigan Center for Integrative Research in Critical Care, University of Michigan
Ann Arbor, 2800 Plymouth Road, Ann Arbor, MI 48109, USA
khoran@gradcenter.cuny.edu
[3] University of York, Heslington, York YO10 5DD, UK

Abstract. We utilize a type of encryption scheme known as a Fully Homomorphic Encryption (FHE) scheme which allows for computation over encrypted data. Our encryption scheme is more efficient than other publicly available FHE schemes, making it more feasible. We conduct simulations based on common scenarios in which this ability is useful. In the first simulation we conduct time series analysis via Recursive Least Squares on both encrypted and unencrypted data and compare the results. In simulation one, it is shown that the error from computing over plaintext data is the same as the error for computing over encrypted data. In the second simulation, we compute two known diagnostic functions over publicly available data in order to calculate computational benchmarks. In simulation two, we see that computation over encrypted data using our method incurs relatively lower costs as compared to a majority of other publicly available methods. By successfully computing over encrypted data we have shown that our FHE scheme permits the use of machine learning algorithms that utilize polynomial kernel functions.

Keywords: Clinical decision support · Data mining ·
Machine learning · Privacy preserving classifier · Encryption

1 Introduction

The field of medical informatics utilizes data mining algorithms that should be built on diverse databases in order to obtain generalizable results [36]. Training, validating, and testing a computational hypothesis typically requires access to large sample datasets that adequately represent any variation within the relevant population [29]. Ideally, data from separate entities, i.e., multiple hospitals and healthcare systems, would be integrated to en-sure an accurate sample. It is evident that healthcare systems vary in location, specialty, and patient care protocols; patients at any particular hospital represent only a subset of the population

© ICST Institute for Computer Sciences, Social Informatics and Telecommunications Engineering 2019
Published by Springer Nature Switzerland AG 2019. All Rights Reserved
A. Compagnoni et al. (Eds.): BICT 2019, LNICST 289, pp. 98–111, 2019.
https://doi.org/10.1007/978-3-030-24202-2_8

and therefore data from a single hospital may contain only a subset of potential illnesses and injuries. Any computational analysis meant to produce diagnostic or prognostic clinical decision support needs to be validated on data collected from multiple healthcare systems. However, ownership and privacy issues greatly limit the possibility of creating such large, comprehensive databases.

Due to the sensitivity of medical data, federally enforced privacy regulations such as Health Insurance Portability and Accountability Act (HIPPA) [7] place firm constraints on data sharing. Therefore, creating or obtaining diversified medical datasets without violating privacy regulations is challenging. Anonymization, the systematic removal of potential patient identifiers, is often seen as a potential solution. However, it has been shown that deanonymization is a relatively simple task. The ease of deanonymization is exemplified by the re-identification of the Governor of Massachusetts within an anonymized health database [11], as well as many other instances. These privacy concerns suggest that medical data should only be released in encrypted format. Thus far, for medical data, there has been a trade-off between utility and security; a useful patient database that allows for computation is insecure, whereas a secure patient database is practically useless for computation and research. This complication stems from the lack of existing technologies that are capable of supporting statistical analysis or machine learning methods on encrypted data. In order to make comprehensive, secure and publicly useful databases possible, the medical community needs to adopt an approach that would allow data analytics on shared encrypted databases while respecting federal privacy restrictions.

The cryptographic community has recently developed fully homomorphic encryption (FHE) schemes, which admit such secure computation. The fully homomorphic property of an encryption function E, can be defined as follows: for any a, b, one has that

$$E(a + b) = E(a) + E(b) \text{ and } E(ab) = E(a)E(b).$$

This implies that for any polynomial function $F(x_1, x_2, \ldots, x_n)$ one has the following property:

$$E(F(x_1, x_2, \ldots, x_n)) = F(E(x_1), E(x_2), \ldots, E(x_n))$$

In other words, computing on an encrypted database yields essentially the same results as computing on an unencrypted database. In this paper we propose a secure machine learning approach based on FHE. This allows researchers and hospitals to run a family of machine learning methods on encrypted data. We offer a solution to the privacy standoff in the case that the desired machine learning algorithm uses polynomial kernel functions.

The FHE scheme most widely accepted as theoretically secure is due to the work of Gentry [14]. This scheme was subsequently improved by Brakerski et al. [5] and the relevant software under development by IBM is available on GitHub [20]. The security of these particular solutions is given by the property of "random self-reducibility". Essentially, finding a solution to the underlying problem is about as hard on average as it is in the worst case. While this property is indeed good evidence of theoretical security, the resulting homomorphic encryption algorithm is too inefficient to be practical.

The conflict can be stated as follows: in order to provide semantic security an encryption algorithm must be randomized, but on the other hand any homomorphism should map zero to zero; an encryption of zero cannot be zero but an encryption of zero must be zero. To resolve this, encryptions of zero are "masked" by "noise" in the aforementioned schemes. The new problem is that during computation on encrypted data this "noise" tends to accumulate and must occasionally be reduced to preserve the correctness of decryption. Therefore, the inefficiency of these schemes can be attributed to the adopted noise management solution. In the above implementations the noise reduction process is *recryption* (also known as bootstrapping), a function that takes a noisy ciphertext and produces an equivalent ciphertext with less noise. Recryption is an expensive procedure and limits both the efficiency and real-life applicability of any existing FHE solution.

There were alternative proposals for FHE following Gentry's, given by Brakerski et al., Ducas et al. and Van Dijk et al., which can be found in their respective works [6,10,35]. While some of these approaches seem conceptually simple and effective, all proposed solutions still involve "bootstrapping". In addition, a majority of the existing methods deal with encrypting Boolean circuits (i.e., AND and OR Boolean operations) as opposed to the "usual" arithmetic operations of multiplication and addition. A "leveled" scheme, an encryption scheme that is typically more efficient than FHE schemes but can only correctly compute a bounded number of operations on ciphertexts, based on adaptations of one of the works above has been developed by Fan and Vercauteren [13], an implementation of this scheme in the R-platform can be found online. In contrast to leveled schemes, our scheme is fully homomorphic and therefore does not require that any parties determine the desired function to compute prior to instantiating the encryption scheme. A further review of FHE schemes and software tools has been provided by Aslett et al. [2].

The first implementation of Gentry's FHE scheme boasted one – albeit large dimensional – bootstrapping operation in 31 min [15]. Subsequent implementations of the improved FHE schemes utilize as benchmarks the computation time for one bootstrap operation as well as computing the AES encryption circuit. The HElib [19], FHEW [10], and TFHE [8] are implementations of alternative encryption schemes. In terms of benchmarking, the fastest and most recent implementation, TFHE, reports 13 ms for the computation of one bootstrapped binary gate.

Recently, secure machine learning techniques have appeared in the literature. Many techniques implement the schemes mentioned above, but many privacy preserving classifiers circumvent FHE and instead make use of leveled homomorphic encryption schemes, garbled circuits, secret sharing, or multi-party computation. For example, Du et al. [9] develop a secure model for linear regression and two-party multivariate classification. This technique does not use FHE, nor does it provide computational benchmarks.

Various techniques attempt to use differential privacy, a tool to protect against database deanonymization, to allow for data sharing. These schemes include those provided by Zhang et al. [38] and Sarwate et al. [32]. These techniques do not implement cryptographic schemes but, if provably secure, could allow for useful patient data warehouses. Yang et al. [37] develop a framework for delegated computation, such as searching encrypted databases in the cloud, which can be used for electronic medical records. This framework, while efficient, does not implement machine learning or FHE.

Graepel et al. [16] construct low-degree polynomial versions of classification algorithms on a leveled homomorphic encryption scheme. Nikolaenko et al. [28] improve upon this performance, reducing computation time, and construct privacy preserving ridge regression using encryption and garbled circuits. This construction does not use an FHE scheme, but is a hybrid construction. Other hybrid constructions exist, such as the constructions that use garbled circuits proposed by Sadeghi et al. [31] and Evans et al. [12] or garbled circuits combined with secret sharing proposed by Lindell et al. [24], but do not implement FHE.

Bos et al. [3] also utilize a leveled homomorphic encryption scheme. The encryption scheme uses a polynomial ring as a platform to allow a client to delegate medical prediction functions to a server. In this setting the algorithm only keeps the patient data private, and allows everyone to know the function.

Bost et al. [4] explore privacy-preserving classifiers in the form of hyperplane decision, Naive Bayes, and decision trees using delegated computation. This is based on additively homomorphic encryption schemes and garbled circuits. Our scheme has the advantage of involving lower communication complexity for a linear classifier. Eventually, our framework could be extended to include more sophisticated learning algorithms such as those addressed in this paper. Many other constructions, such as that by Aono et al. [1], merely use an additively homomorphic encryption scheme for computations such as logistic regression or statistical analysis [21,23,27].

Liu et al. [25] use the IBM implementation of FHE to delegate the computation of Support Vector Machine classification on encrypted data to a server. The paper provides benchmarks for encryption and decryption time, as well as homomorphic multiplication and addition, on a small dataset. Our scheme, as seen in our results, greatly reduces computation time. Our approach to homomorphic encryption is based on the use of mathematical rings and homomorphisms between those rings. This framework has the additional benefit of avoiding any computational overhead due to converting between "real-life arithmetic" and Boolean circuits. In this paper we avoid detailing the specifics of our encryption method, which are outlined in [17,22]. We will say that since the ciphertext ring, in which computations on encrypted data are performed, has a very simple structure all computations within our scheme are orders of magnitude more efficient than the schemes mentioned.

The advantage of using this specific FHE scheme is error-less computation with guaranteed security, in contrast to some of the non-FHE solutions provided above. The main contribution of this paper is that our FHE scheme is a fully homomorphic scheme that is efficient enough to allow for machine learning on encrypted data. We provide simulations, error, and computational overhead of using our scheme for this purpose. While the works in this paper are generally applicable to a multitude of scenarios, we consider two specific applications for machine learning over encrypted data. These two scenarios will be simulated in the following experiments. Most of the data utilized is actual patient data; the term *simulation* refers to the simulation of both the communication between parties, and computations performed by each party.

In the first scenario, a hospital $\mathcal{H}$ has a private database of patients' data, D, which cannot be shared with an external entity. On the other hand, a research center $\mathcal{C}$ would like to use D to construct a machine learning model, for example to build a function F to calculate a risk assessment score, or perform actuarial analyses. In practice, $\mathcal{C}$ could be a research center, another hospital, a pharmaceutical company, an insurance company, or any other third party. Additionally, $\mathcal{C}$ may be reluctant to share its intellectual property, F, with $\mathcal{H}$. This scenario happens quite often within the research community. Researchers find it difficult to train generalizable models because of the lack of public data. This scenario is, in part, simulated in this paper, when we compare the error results of training a function on encrypted and unencrypted data.

In the second scenario, $\mathcal{C}$ has a collection of multivariate functions

$$F_i : X_n \to Y \text{ for } 1 \leq i \leq k$$

representing a specific machine learning algorithm, while $\mathcal{H}$ has the inputs representing the medical data for patient $x = (x_1, x_2, \ldots, x_n)$. The set of known functions, $F = F_{i}{}_{i=1}^{k}$ are expected to predict a diagnosis or prognosis for x, for example the chances of patient x having Parkinson's, cancer, heart attack, etc. The difference between this second scenario and the first outlined scenario is that in this scenario we assume that F is a known function (e.g., we know how to calculate Glasgow Coma Scale (GCS) using eye, verbal, and motor responses). Our goal is to apply F on both encrypted and unencrypted data and show that the output of a known function evaluation on both sets of data is the same. Outputs of these functions could be "health metrics", "severity scores" and other clinical functions typically computed as a linear combination of privacy-protected factors, such as quantitative clinical or physiological patient data with a set of weights (coefficients). In practice, these functions are often designed to receive integer values as input variables and produce an integer as an output score. There are a large number of such studies conducted for diseases such as diabetes [33]. In many such modeling tasks, particularly when designing these models as commercial products, it is highly desirable to design, test and validate the functions privately. This scenario, in part, is simulated in this paper, where we apply known classification functions to encrypted and unencrypted data.

In very general terms, a potential exchange between the hospital $\mathcal{H}$ and a center $\mathcal{C}$ goes as follows. $\mathcal{H}$ encrypts data, $x_1, x_2, \ldots, x_n$, of a particular patient x with E and sends the encrypted values $E(x_1), E(x_2), \ldots, E(x_n)$ to $\mathcal{C}$. $\mathcal{C}$ applies the private function F to the encrypted data, thereby computing

$F(E(x_1), E(x_2), \ldots, E(x_n))$. The result of this computation, by the fully homomorphic property of E, is equal to $E(F(x_1, x_2, \ldots, x_n))$. Next $\mathcal{C}$ sends this result to $\mathcal{H}$, who decrypts the value and thus recovers $F(x_1, x_2, \ldots, x_n)$. $\mathcal{H}$ then sends this decrypted value back to $\mathcal{C}$. Based on the received evaluation of F, the final plaintext message, the hospital $\mathcal{H}$ has a diagnosis or risk assessment for patient x. Thus, $\mathcal{C}$ never learns the plaintext patient data x and $\mathcal{H}$ never learns the function F, but does learn $F(x)$.

2 Encryption Overview

We first give a general description of FHE and its relevant terminology. The term *plaintext* refers to unencrypted data, whereas *ciphertext* refers to encrypted data. For our applications all data will be integers. Additionally, in order to enable the encryption process to select elements of the plaintext space uniformly at random, we require that the plaintext space be finite. This randomness is required for security. In the case that actual medical data is not measured in integers, we can merely "re-scale" the measurement by multiplying the value with a sufficiently large integer, to ensure that this property holds. Thus, we can guarantee that the set of plaintexts, $\mathbb{Z}_p$, will be the ring of integers modulo p. The prime number p here should be large enough that all plaintexts would be integers much less than p. Specific implementation details can be found in supplementary materials, [17,22]. To keep this paper relatively self-contained we describe our general encryption scheme below. Although certain details of the scheme are provided, the FHE scheme will essentially be implemented as a black box encryption function for the duration of this work.

- Plaintexts are elements of the ring $\mathbb{Z}_p$. We start by embedding $\mathbb{Z}_p$ into a larger ring R, which is a direct sum of several copies of $\mathbb{Z}_p$. We denote this embedding by α and the inverse map by β. The ring R can be public but both α and β are private; α is a part of the private encryption key, while β is a part of the private decryption key.
- Ciphertexts are elements of a ring S, such that $R \subset S$ is a subring of S. In our scheme S is, again, a direct sum of several copies of $\mathbb{Z}_p$. The ring S is public but contains a private ideal I such that $S/I = R$.
- Encryption is given by $E(u) = u + E(0)$, for an element $u \in R$, where $E(0)$ is a random element of the private ideal I. This encryption function is a homomorphism; the additive property is clear. For multiplication we have, for some $j_1, j_2, j_3 \in I$, that

$$E(u)E(v) = (u + j_1)(v + j_2) = uv + j_1 u + u j_2 + j_1 j_2 = uv + j_3 = E(uv)$$

- Decryption is computed with a private decryption key, a map $\rho : S \to R$ that takes every element of the ideal I to 0, followed by the map β from R to $\mathbb{Z}_p$.

Here is a diagram to visualize our general scheme:

$$\mathbb{Z}_p \xrightarrow{\alpha} R \xrightarrow{E} S \xrightarrow{\rho} R \xrightarrow{\beta} \mathbb{Z}_p$$

It should be noted that the number of distinct fully homomorphic embeddings, α, from $\mathbb{Z}_p$ into R is quite large, even if R is a direct sum of only a few copies of $\mathbb{Z}_p$. To explain this we first mention that a map $\alpha : 1 \to \alpha(1)$ extends to an embedding of $\mathbb{Z}_p$ into R if and only if $\alpha(1)$ is an idempotent of R, i.e., $\alpha(1)^2 = \alpha(1)$. When R is a direct sum of n copies of $\mathbb{Z}_p$, then it has 2^n idempotents, so there are 2^n different embeddings of $\mathbb{Z}_p$ into R. Similarly, a fully homomorphic encryption function $E : R \to S$ is an embedding of rings. Therefore, a similar argument applies to counting embeddings of R into S.

From the security point of view, it is known that all FHE schemes have a theoretical vulnerability [18]. In our scheme, the ideal I used for encryptions of zero is finite dimensional and therefore accumulating sufficiently many encryptions of zero may give an adversary the ability to recover I. Fortunately, the recovery of I is not a security threat from a practical point of view because I is not a part of the decryption key; the decryption key consists of the maps ρ and β. If an adversary has recovered the ideal I they may be able to recognize all encryptions of zero, but in real-life scenarios encryptions of zero are not frequently transmitted. We note that if a plaintext is "close" to zero, which can be the case in a medical database, the corresponding ciphertext does not have to be "close" to zero; the proposed encryption function does not preserve any metric and therefore does not preserve distance from zero. In fact, it is easy to see that any encryption function which does preserve such a metric cannot possibly be secure.

Our encryption scheme is completely secure against Ciphertext-Only Attack (COA). This means that a "hacker" who retrieves any encrypted portion of a database, regardless of the size, has only a negligible probability of correctly decrypting any portion of the database. COA security is fairly easy to achieve with a private-key non-homomorphic encryption. However, this property becomes very nontrivial for FHE. One of the main reasons we are able to achieve this property with our FHE scheme is the incredibly large number of fully homomorphic embeddings of $\mathbb{Z}_p$ in our public ring S. This implies that there are many different ways to decrypt any given ciphertext, but only one decryption is correct. Therefore, the probability to decrypt correctly is $\frac{1}{M}$, where M is the total number of possible decryptions. If M is large enough this probability is negligible. Our suggested parameters, which yield this security guarantee, have M on the order of 2^{128}.

3   Simulations

A machine learning solution should not only calculate the essential functions on encrypted patient data but also preserve privacy. In this paper, we focus on publicly available databases and high impact functions. First, we have conducted time series analysis on both synthetic and real heart rate data. Second, we have chosen to calculate known functions on Diabetes data, as well as a known predictive function on Parkinson's data. To illustrate the correctness and efficiency of our FHE scheme we perform experiments on this data. We report both error and computational overhead in the Results Section of the paper.

In the first simulation we apply a function, e.g., develop a model, on both encrypted and plain-text datasets to estimate/predict an output for each dataset and show that there is no difference between the outputs. We perform linear time-series analysis and fit a model to the encrypted data to understand the underlying structure and perform forecasting, monitoring and so on. Here, we use Recursive Least Squared (RLS) as the F function.

In the second simulation we apply a known classification function on encrypted data and measure the computational overhead. We focus on publicly available medical data for Parkinson's and diabetes, in order to illustrate the efficacy of our scheme on medical data.

3.1   Experiment One

In this simulation we play the role of an external entity C: we are given an encrypted database D and train a function F on the encrypted data. As an example, an assessment function can tell whether a patient has a risk of heart attack, based on his/her medical data such as age, blood pressure, heart rate, etc. With the proposed encryption method, we can obtain an assessment function from encrypted data when we are given the relevant formula for obtaining that function from unencrypted data.

We perform linear univariate time series analysis on two separate databases using the RLS algorithm. Time series data can allow for high level medical analysis, including features calculated from variations in heart rate, heart rate variability, blood pressure, etc. The RLS algorithm implemented in this section is fairly straightforward.

Synthetic Data. The first database (DB 1) consists of synthetic data, generated according to a known distribution. The input signal is $x(n)$ and we generated the output signal, $y(n)$, via an equation presented in [26]:

$$y(n) = \sum_{k=0}^{N} h(k)x(n-k)$$

where $h(k) = 1/(k+1)$. Note that these parameter values were chosen arbitrarily: any values of N would yield similar results. We have applied the RLS algorithm to both the encrypted and unencrypted data. We then measured the mean squared error between the actual, known function value and our estimated function value after the 1000 iterations. The goal, of course, is to have minimal error, thereby indicating the feasibility of running the RLS function on encrypted data. Table 1 contains the error results for this database as well as the time difference between plaintext and encrypted data computations for one output.

Santa Fe Time Series Data. Our second database (DB 2) was time-series data from the Santa Fe Time Series Competition [30]. For simplicity, only Heart

Rate Variation (HRV) over time was considered. We applied RLS to both the unencrypted and encrypted heart rate data of patients. We measured the error between the actual heart rate and the one predicted by our function on both unencrypted and encrypted data. To reiterate, the error between our function's prediction and actual heart rate is reported in both situations. Because this simulation involves actual prediction, we hope for the error values on encrypted and unencrypted data to be as close as possible. The given measurements of the heart rate were in the range $[70.00, 100.00]$. Table One contains the results of our tests.

3.2   Experiment Two

In this simulation we have run several statistical tests on real-life databases encrypted by our method, including diabetic data [33] and Parkinson's data [34]. This simulation has real-life applications. For example, a hospital $\mathcal{H}$ has a private database D of patients' medical/clinical data. At the same time, a research center $\mathcal{C}$ has statistical tools that could help $\mathcal{H}$ assess risk. We therefore calculate this classification function on the following databases while maintaining the privacy of both the model and the data. Note that this scenario does not involve any training, merely homomorphic function evaluation on ciphertexts. We compute the same known polynomial function on encrypted and unencrypted data. We measured the time for encryption and decryption of the data for one patient as well as for the whole database. The execution time of a linear function was captured for each encrypted dataset. Table 2 provides the results of simulation.

Diabetic Database. The diabetic data (D) represents 10 years (1999–2008) of clinical care at 130 US hospitals and integrated delivery networks. The publically available input data contains many parameters on over $70,000$ patients, including demographic and clinical values. Each patients data includes over 50 features, representing patient and hospital outcomes. The predetermined classification function takes these recorded values, such as race, age and time of stay in hospital, and outputs a re-admission prediction.

Parkinson's Database. The Parkinson's data (P) encompasses a range of biomedical voice measurements from 42 patients with early-stage Parkinson's disease, recruited for a six-month trial of remote symptom progression monitoring. These recordings were automatically captured in the patients' homes. Parameters include demographic and clinical data, as well as features calculated from the recordings such as jitter and shimmer. The known function under consideration takes the patient data, including the dysphonia measurements, as input and out- puts the Pitch Period Entropy (PPE). The PPE value is highly correlated with the progression of Parkinson's disease. The fixed function under consideration is a linear combination of certain attributes in patient vector $x = \langle x_0, x_1, x_2, \ldots, x_n \rangle$, with precomputed coefficients $a_k : 1 \leq k \leq n$ taking the following form:

$$\text{PPE} = x_0 + \sum_{k=1}^{18} a_k x_k$$

4   Results

Tables 1 and 2 report the findings of the simulations. Table 1 reports the size of the RLS window, N, the computation error over plain-text and encrypted data, as well as the difference between these errors and the computational time. Table 2 reports the size of each database, the time to encrypt and decrypt both a single patient and the database, as well as the cost of function evaluation.

Table 1. Experiment One: RLS on Ciphertexts and Plaintexts. This table provides a comparison of RLS applied on both databases, over encrypted and unencrypted data

Database	N	Plaintext error	Encrypted error	Error difference	Time
DB 1	3	3.26×10^{-4}	3.29×10^{-4}	3×10^{-6}	0.004 ms
DB 1	9	2.95×10^{-4}	3.02×10^{-4}	7×10^{-6}	0.009 ms
DB 2	3	1.264	1.269	0.005	0.005 ms
DB 2	9	1.124	1.129	0.005	0.01 ms

In this scope, the importance of these results lies not in the error values themselves but in the proximity between the error values. If the error values are close, this implies that RLS behaves nearly identically on ciphertexts and plaintexts. The closer the error values are the less error can be attributed to the encryption scheme. Ideally, these error values would be the same but, because we must scale all plaintext values to integers prior to encryption, we introduce rounding error. In other words, the difference between the encrypted and unencrypted error values is due to rounding error generated by rescaling the real numbers to integer values. The column "Encrypted Error" in Table 1 reports the error from the recursive function. This computation includes embedding all data into the integers modulo p via a common scaling integer, a power of ten, encrypting, computing, decrypting, and dividing by the scaling integer.

Clearly, the accuracy of the assessment function on the synthetic data is practically the same for unencrypted and encrypted data. This shows the ability of the encryption function to handle RLS training while maintaining correctness. It can be seen that the difference in error over encrypted and unencrypted data for both the synthetic and real situations is negligible. This illustrates that basic machine learning models, those that utilize the proper arithmetic operations, can be trained on ciphertexts while maintaining correctness. Time series analysis is a fundamental aspect of medical data analytics and these experiments have shown that our FHE scheme permits the necessary computations for such analytics.

Note that the difference in error does not scale with N, the RLS window size. This shows that the error introduced by increasing the number of variables involved in computation is not expected to generate significant error. Unavoidably, encrypted computation time is larger than unencrypted computation time. While this computational overhead will increase with the number of patients, this scheme is feasible for performing computation on encrypted data. Table One shows that computation time with our method is still quite practical. This is not the case with alternative FHE implementation methods. See, for example, the work done by Ducas and Micciancio [10], which achieves a single bootstrapped NAND computation in 0.69 s. Additionally, work provided by Aslett et al. [2] claim a single scalar addition at 0.003 s and a single scalar multiplication at 0.084 s even using high performance computers.

Table 2. Experiment Two: Known functions computed on publicly available data.

DB	DB size MB	Patient records	Time to encrypt record	Time to encrypt DB	Time to decrypt record	Time to decrypt DB	Time to evaluate function
D	19	101767	0.02445	2488	0.18152	18472.7	1×10^{-5}
P	1	5876	0.01194	70.1	0.09798	575.7	6×10^{-6}

The results of simulation two can be seen in Table 2. With this simulation we showed that given a polynomial function, e.g., a medical diagnostic or prognostic function, both H and C can evaluate the function homomorphically on encrypted data. This allows C to maintain the integrity of its intellectual property, F, while still using the function to classify private data D. There is no discrepancy between the assessment, i.e., function evaluation, on each individual patient regardless of whether the function was calculated on encrypted or unencrypted data. Therefore, instead of reporting any error measurement we report the efficiency of the scheme and computation on ciphertexts by applying the same linear functions to unencrypted and encrypted data. The efficiency of the scheme is noteworthy - with this simulation we have shown that function evaluation can be performed relatively quickly using our FHE scheme. In this table it is also notable that the time to decrypt is higher than the time to encrypt. This discrepancy is due to the fact that encryption is essentially an addition, while decryption requires computing the remainder of a ciphertext modulo the ideal I via the decryption map.

5   Conclusions

We applied our method of Fully Homomorphic Encryption (FHE) to calculate medical diagnostic functions based on encrypted medical data. Our FHE scheme permits the use of machine learning algorithms that utilize polynomial kernel functions. These computations allow for medical diagnostics to be performed on encrypted data, maintaining the privacy of patient data. We outlined example scenarios where secure machine learning could be useful within the medical community, considering the protection of patient data as well as a researchers intellectual property.

Time series analysis was performed on synthetic and real data, showing that the increase in error is negligible when operating on encrypted data. Then known classification functions were applied to public medical databases without introducing additional error that illustrates the efficiency of our encryption method.

We have shown that it is possible to train an assessment function on encrypted data provided that relevant formulas for obtaining such a function from unencrypted data are available. Our method provides very efficient computation on encrypted data, which allows us to compute any polynomial function on a single patient's encrypted data in a fraction of a second. A polynomial function can be computed on a medical encrypted database in a couple of minutes.

We have shown that the encryption scheme is efficient enough to be practical, secure and correct for linear classifiers as well as time series analysis. Theoretically, it is possible to extend the applicability of the scheme to different families of machine learning functions. Planned future work includes adapting the current encryption scheme to working with non-linear classifiers, such as Volterra systems. This generalization would show that the encryption scheme permits a larger class of machine learning functions to be computed on encrypted data.

The above implementation of the FHE function allows for efficient data mining without decryption while maintaining correctness. Therefore, it is completely feasible to consider the utilization of this encryption function for highly sensitive, private, and federally regulated medical data.

References

1. Aono, Y., Hayashi, T., Trieu Phong, L., Wang, L.: Scalable and secure logistic regression via homomorphic encryption. In: Proceedings of the Sixth ACM Conference on Data and Application Security and Privacy, pp. 142–144. ACM (2016)
2. Aslett, L.J., Esperança, P.M., Holmes, C.C.: A review of homomorphic encryption and software tools for encrypted statistical machine learning. arXiv preprint arXiv:1508.06574 (2015)
3. Bos, J.W., Lauter, K., Naehrig, M.: Private predictive analysis on encrypted medical data. J. Biomed. Inform. **50**, 234–243 (2014). Special Issue on Informatics Methods in Medical Privacy
4. Bost, R., Popa, R.A., Tu, S., Goldwasser, S.: Machine learning classification over encrypted data. In: NDSS (2015)
5. Brakerski, Z., Gentry, C., Vaikuntanathan, V.: (Leveled) fully homomorphic encryption without bootstrapping. ACM Trans. Computat. Theor. (TOCT) **6**(3), 13 (2014)
6. Brakerski, Z., Vaikuntanathan, V.: Efficient fully homomorphic encryption from (standard) LWE. In: Proceedings of the 2011 IEEE 52nd Annual Symposium on Foundations of Computer Science, FOCS 2011, pp. 97–106. IEEE Computer Society, Washington, DC (2011)

7. Centers for Disease Control and Prevention: HIPAA privacy rule and public health. Guidance from CDC and the US department of health and human services. MMWR Morb. Mortal. Wkly. Rep. **52**(Suppl. 1), 1–17 (2003)

8. Chillotti, I., Gama, N., Georgieva, M., Izabachène, M.: Faster fully homomorphic encryption: bootstrapping in less than 0.1 seconds. In: Cheon, J.H., Takagi, T. (eds.) ASIACRYPT 2016. LNCS, vol. 10031, pp. 3–33. Springer, Heidelberg (2016). https://doi.org/10.1007/978-3-662-53887-6_1

9. Du, W., Han, Y.S., Chen, S.: Privacy-preserving multivariate statistical analysis: linear regression and classification. In: Proceedings of the 2004 SIAM International Conference on Data Mining, pp. 222–233. SIAM (2004)

10. Ducas, L., Micciancio, D.: FHEW: bootstrapping homomorphic encryption in less than a second. In: Oswald, E., Fischlin, M. (eds.) EUROCRYPT 2015. LNCS, vol. 9056, pp. 617–640. Springer, Heidelberg (2015). https://doi.org/10.1007/978-3-662-46800-5_24

11. El Emam, K., Jonker, E., Arbuckle, L., Malin, B.: A systematic review of re-identification attacks on health data. PloS One **6**(12), e28071 (2011)

12. Evans, D., Huang, Y., Katz, J., Malka, L.: Efficient privacy-preserving biometric identification. In: Proceedings of the 17th conference Network and Distributed System Security Symposium, NDSS (2011)

13. Fan, J., Vercauteren, F.: Somewhat practical fully homomorphic encryption. IACR Cryptology ePrint Archive 2012, 144 (2012)

14. Gentry, C., Boneh, D.: A Fully Homomorphic Encryption Scheme, vol. 20. Stanford University, Stanford (2009)

15. Gentry, C., Halevi, S.: Implementing Gentry's fully-homomorphic encryption scheme. In: Paterson, K.G. (ed.) EUROCRYPT 2011. LNCS, vol. 6632, pp. 129–148. Springer, Heidelberg (2011). https://doi.org/10.1007/978-3-642-20465-4_9

16. Graepel, T., Lauter, K., Naehrig, M.: ML confidential: machine learning on encrypted data. In: Kwon, T., Lee, M.-K., Kwon, D. (eds.) ICISC 2012. LNCS, vol. 7839, pp. 1–21. Springer, Heidelberg (2013). https://doi.org/10.1007/978-3-642-37682-5_1

17. Gribov, A., Kahrobaei, D., Shpilrain, V.: Practical private-key fully homomorphic encryption in rings. Groups Complex. Cryptol. **10**(1), 17–27 (2018)

18. Grigoriev, D., Ponomarenko, I.: Homomorphic public-key cryptosystems over groups and rings. arXiv preprint cs/0309010 (2003)

19. Halevi, S., Shoup, V.: Algorithms in HElib. In: Garay, J.A., Gennaro, R. (eds.) CRYPTO 2014. LNCS, vol. 8616, pp. 554–571. Springer, Heidelberg (2014). https://doi.org/10.1007/978-3-662-44371-2_31

20. Halevi, S., Shoup, V.: Helib (2014). HELib: https://github.com.shaih/HElib

21. Hall, R., Fienberg, S.E., Nardi, Y.: Secure multiple linear regression based on homomorphic encryption. J. Off. Stat. **27**(4), 669 (2011)

22. Kahrobaei, D., Lam, H., Shpilrain, V.: System and method for private-key fully homomorphic encryption and private search between rings. Patent US20170063526, 25 August 2017

23. Lauter, K., López-Alt, A., Naehrig, M.: Private computation on encrypted genomic data. In: Aranha, D.F., Menezes, A. (eds.) LATINCRYPT 2014. LNCS, vol. 8895, pp. 3–27. Springer, Cham (2015). https://doi.org/10.1007/978-3-319-16295-9_1

24. Lindell, P.: Privacy preserving data mining. J. Cryptol. **15**(3), 177–206 (2002)

25. Liu, F., Ng, W.K., Zhang, W.: Encrypted SVM for outsourced data mining. In: 2015 IEEE 8th International Conference on Cloud Computing (CLOUD), pp. 1085–1092. IEEE (2015)

26. Ljung, L.: System Identification: Theory for the User. Prentice-Hall, Upper Saddle River (1987)
27. Naehrig, M., Lauter, K., Vaikuntanathan, V.: Can homomorphic encryption be practical? In: Proceedings of the 3rd ACM Workshop on Cloud Computing Security Workshop, pp. 113–124. ACM (2011)
28. Nikolaenko, V., Weinsberg, U., Ioannidis, S., Joye, M., Boneh, D., Taft, N.: Privacy-preserving ridge regression on hundreds of millions of records. In: 2013 IEEE Symposium on Security and Privacy, pp. 334–348, May 2013
29. Poggio, T., Smale, S.: The mathematics of learning: dealing with data. Not. AMS **50**(5), 537–544 (2003)
30. Rigney, D.R., Goldberger, A.L., Ocasio, W., Ichimaru, Y., Moody, G.B., Mark, R.: Multi-channel physiological data: description and analysis (1993)
31. Sadeghi, A.-R., Schneider, T., Wehrenberg, I.: Efficient privacy-preserving face recognition. In: Lee, D., Hong, S. (eds.) ICISC 2009. LNCS, vol. 5984, pp. 229–244. Springer, Heidelberg (2010). https://doi.org/10.1007/978-3-642-14423-3_16
32. Sarwate, A.D., Chaudhuri, K.: Signal processing and machine learning with differential privacy: algorithms and challenges for continuous data. IEEESignal Process. Mag. **30**(5), 86–94 (2013)
33. Strack, B., et al.: Impact of HbA1c measurement on hospital readmission rates: analysis of 70,000 clinical database patient records. BioMed Res. Int. **2014** (2014)
34. Tsanas, A., Little, M.A., McSharry, P.E., Ramig, L.O.: Accurate telemonitoring of Parkinson's disease progression by noninvasive speech tests. IEEE Trans. Biomed. Eng. **57**(4), 884–893 (2010)
35. van Dijk, M., Gentry, C., Halevi, S., Vaikuntanathan, V.: Fully homomorphic encryption over the integers. In: Gilbert, H. (ed.) EUROCRYPT 2010. LNCS, vol. 6110, pp. 24–43. Springer, Heidelberg (2010). https://doi.org/10.1007/978-3-642-13190-5_2
36. Wiens, J., Guttag, J., Horvitz, E.: A study in transfer learning: leveraging data from multiple hospitals to enhance hospital-specific predictions. J. Am. Med. Inform. Assoc. **21**(4), 699–706 (2014)
37. Yang, J.J., Li, J.Q., Niu, Y.: A hybrid solution for privacy preserving medical data sharing in the cloud environment. Futur. Gener. Comput. Syst. **43**, 74–86 (2015)
38. Zhang, J., Zhang, Z., Xiao, X., Yang, Y., Winslett, M.: Functional mechanism: regression analysis under differential privacy. Proc. VLDB Endow. **5**(11), 1364–1375 (2012)

Evolutionary Multi-objective Optimization for Evolving Soft Robots in Different Environments

Jun Ogawa[✉][iD]

Yamagata University, 4-3-16 Jonan, Yonezawa, Yamagata, Japan
jun.ogawa@yz.yamagata-u.ac.jp
http://www.junogawa.com/

Abstract. Evolutionary robotics is an approach for optimizing a robotic control system and structure based on the bio-inspired mechanism of adaptiogenesis. Conventional evolutionary robotics assigns a task and an evaluation to a virtual robot and acquires an optimal control system. In many cases, however, the robot is composed of a few rigid primitives and the morphology imitates that of real animals, insects, and artifacts. This paper proposes a novel approach to evolutionary robotics combining morphological evolution and soft robotics to optimize the control system of a soft robot. Our method calculates the relational dynamics among morphological changes and autonomous behavior for neuro-evolution (NE) with the development of a complex soft-bodied robot and the accomplishment of multiple tasks. We develop a soft-bodied robot composed of heterogeneous materials in two stages: a development stage and a locomotion stage, and we optimize these robotic structures by combining an artificial neural network (ANN) and age-fitness pareto optimization (AFP). These body structures of the robot are determined depending on three genetic rules and some voxels for evolving the ANN. In terms of our experimental results, our approach enabled us to develop some adaptive structural robots that simultaneously acquire behavior for crawling both on the ground and underwater. Subsequently, we discovered an unintentional morphology and behavior (e.g., walking, swimming, and crawling) of the soft robot through the evolutionary process. Some of the robots have high generalization ability with the ability to crawl to any target in any direction by only learning a one-directional crawling task.

Keywords: Evolutionary robotics · Soft robotics · Neural network · Pareto optimization · Artificial life

1 Introduction

For most evolutionary robots, the controller and morphology, which is inspired by real animals and insects, are given in advance [1–3]. The robot morphologies include some biases by human recognition, however, these structures of animals and insects depend on what they learn during the development, and in turn

© ICST Institute for Computer Sciences, Social Informatics and Telecommunications Engineering 2019
Published by Springer Nature Switzerland AG 2019. All Rights Reserved
A. Compagnoni et al. (Eds.): BICT 2019, LNICST 289, pp. 112–131, 2019.
https://doi.org/10.1007/978-3-030-24202-2_9

the structure decides how they learn. In spite of the difference between human imagination and a real evolutionary result, these biases involuntarily associate the morphologies with specific behavior based on human experiences. Obtaining a more unintentional and adaptive evolutionary robot is one of the significant challenges in the field of evolutionary robotics, and it has been studied actively in areas such as artificial life, morphological evolution and computer graphics, and animation [4–13].

Morphogenesis engineering fabricates a robot morphology based on the mechanism of self-organization in natural systems, including the development of intelligence and composition of heterogeneous components. Sims demonstrated a virtual robot that simultaneously evolves a neuro-controller and morphology, and contributed hugely to the field of robotics [14]. This robot, however, was largely problematic in that in cases in which the robot is built by simple rigid components under simple developmental rules defined in advance, the optimal robot would have almost the same form and would be unable to adapt to different environments and multiple tasks. Doursat et al. proposed a way to design virtual soft-bodied robots by growing fine-gained multicellular [15, 16]. This study provides two major rules – cell adhesion and cell division – into each spherical cellular shape. A pair of two cells receives a force based on three conditions depending on the distance between them. The results of their work showed that the robot generated four limb-like parts in the body and it acquired the ability to perform two tasks: rolling a rigid sphere and walking to a place located upstairs. They provided the potential properties that each cell has a certain kind of body part such a right limb, left hand, or short length in order to grow a structure such as that of real creatures. Joachimczak et al. proposed artificial metamorphosis as a method of evolving self-reconfiguring soft-bodied robots from the viewpoint of evolution from a tadpole to a frog [17–19]. They created a robot with reduced human biases as much as possible in order to understand the evolutionary process in real creatures, and they adopted a straightforward approach by combining mere neuro-evolution method and propagation mechanism of virtual morphogens for metamorphosis. As the main result they showed that these robots evolved from fish-like creatures to bipedal creatures and ascertained that the structures of some real creatures are adaptive for moving from water to land or from land to water.

The result verified the above-mentioned wonderful findings through two-dimensional creatures; however, the research would need to focus on three-dimensional creatures such as real creatures to obtain more advanced results. In the field of computer graphics (CG) and physical animation, Geijtenbeek et al. proposed an optimization method for obtaining flexible muscle-based bipedal robots in computer simulation [20]. The approach is to optimize the place at which the virtual muscle fibers are connected to two rigid boxes and they obtained several musculo-skeletal robots that can perform more flexible and animal-like motions than robots obtained via conventional approaches. The important point in this study is that there are unexpected effective structures for improving behavioral awkwardness to obtain a desired motion. This indicates that a human-designed structure will not always be effective for an assumed task. Thus, we would need to dispose of such biases by human recognition to truly obtain a robot with an adaptive morphology to achieve a given task because there is no guarantee that every designed robot would have the ability to acquire such a skill. We believe the coupled dynamics of structure development and learning behavior of a given task would enable the construction of more complex and unintentional structures. The complicated structure obtained through their evolutionary experience of the development can be more robust and adaptive and it might be able to shed light on a different perspective of the life structure. Meanwhile, soft-bodied robots generally have higher robustness and adaptation than rigid-bodied robots because these robots can deform themselves [21,22]. In the field of evolutionary robotics, controlling the physical behavior of those soft robots is also important and challenging because the controller of a soft robot needs to control the body actuators considering their structural deformation acquired from the surrounding environmental effects.

We pursue the ultimate objective of establishing a novel way to evolve more robust and adaptive soft-bodied robots in different environments, with multiple tasks, while the robot is simultaneously evolving its morphology and intelligence. This paper discusses how to design an evolutionary strategy and simulation foundation for considering the less-biased development of soft-bodied robots in different environments. Our simulation model considers a robot structure that consists of heterogeneous materials, which enables us to suggest an embryogenesis mechanism based on physiology. We then compare the results of locomotion experiments in which soft-reconfiguration soft-bodied robots evolve on the ground and underwater to acquire the behavior to crawl around their environments, and we analyze an adaptation for these evolved robots.

This paper makes the following contributions: First, we establish a novel way to model a soft-bodied robot with a coupling of dynamics between morphological development and behavioral learning by using an artificial neural network. In particular, our work proposes a novel evolutionary strategy that develops the structure of the robot and simultaneously evaluates multiple types of behavior in different environments. In this way it would be able to evolve some robots considering each environmental constraint, such that they gradually adapt their behavior to these environments. Then, we discuss the way in which the morphology evolves and what the robot learns on these constraints evaluating robot morphologies and behavior to achieve multiple tasks in these environments.

2   Methods

2.1   Approach

In the following section we describe the essential concept of our development system for evolving soft-bodied robots. Here, there is a gene regulatory network (GRN) as a simple description of controlling such cellular behavior [23].

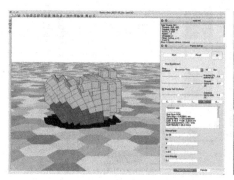 

Fig. 1. (Left) A robot morphology using four material voxels developed by Voxelyze. (Right) Example of automatic design and a fabricated soft-bodied robot by Voxelyze with silicone rubber (Color figure online)

The network dominates the extent to which the fate of every biological cell is determined. We adopt an ANN to introduce the concept of GRN into our developmental system, and attempt to evolve the ANN through physical simulation that has two stages – development and locomotion – for obtaining an adaptive soft-bodied robot. We use some voxels to create robot morphology, and each voxel has the role of an actuator such as muscle or static support such as bone or fat in the general body composition of animals. One part of the simulation is the development stage in which the robot receives a signal from the network and changes its own structure with voxel material. Another part of the simulation is the locomotion stage in which the designed robot, which is obtained at the end of the development stage, moves around in a given environment in advance for a fixed period, and this stage is subsequently executed after the development stage. Our approach is not to create a central control system that manages the dynamics of all voxels in detail in order to achieve distributed control in the soft-bodied robot by using only local interaction among multiple actuators without any specifications. Besides, we never determine where every voxel is placed in advance to reduce the human bias of development as much as possible. Namely, we allow for a feed-forward ANN that develops the robot morphology and the actuator properties, and the ANN begins developing from a single voxel to design a completed robot. We believe that the straightforward approach is able to provide evolution-ranges toward more adaptive and unexpected robots. We assume that our robot is developed through sequential voxel addition and deletion. Moreover, we provide one of four material properties – hard muscle, soft muscle, bone, and fat – to all voxels based on the output of the ANN because we focus on making a model of a real-like creature as a soft-bodied robot.

2.2   Voxelyze

In 2013, Hiller et al. developed computer software named Voxelyze [24–27], for the physical simulation of a soft-bodied robot composed of material voxels. Voxelyze provides several properties – size, Young modulus, density, coefficient of

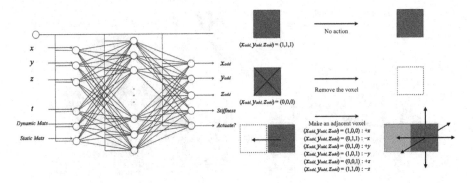

Fig. 2. (Left) The artificial neural network for making the morphology of a soft-bodied robot with homogeneous materials. (Right) The three developmental rules, which are no action, removing the voxel, and adding a new voxel.

thermal expansion, friction, and damper – for insertion into the voxel, and it can verify the control system of a more complex soft-bodied robot.

Additionally, the morphology can be constructed by using a 3D printer. Figure 1 shows a sample robot created by Voxelyze and a 3D printed robot in the real world based on Voxelyze. We define four materials for use in the voxel. The first material is a hard muscle voxel (red voxel) that has high Young modulus. The second material is a soft muscle voxel (orange voxel) that has a lower Young modulus than the hard muscle. Those voxels work as dynamic actuators in the robot body and periodically vibrate depending on a sine function. The third material is bone voxel (blue voxel) and is a static support object with the same Young modulus as the hard muscle.

The fourth material is a fat voxel (cyan voxel), and it is also a static support object with a lower Young modulus than the bone voxel. Our modeling constructs an all-connected voxel as a robot in 11 × 11 × 11 design space that is able to place each voxel.

2.3 Evolving Artificial Neural Network

The ANN is a well-known brain model and consists of a set of neurons and synapses. Our network is composed of six input neurons and five output neurons (see Fig. 2). We provide a time signal that merely increases in [0:1] and the three-dimensional position – in x-, y-, and z-coordinates – as inputs to the ANN. Additionally, we introduce two morphogen neurons into the ANN, and the ANN can consider interaction among adjacent voxels by corresponding to two virtual morphogens, namely the rate of dynamic voxels (hard and soft muscle) and the rate of static voxels (bone and fat) in the surrounding voxels. Figure 2 shows the structure of the ANN and the construction rule of voxels by the ANN. There are two types of output neuron to determine which direction creates the new voxel and which material defines the voxel. For one type of three output neurons (xadd, yadd and zadd) shown in Fig. 2, the ANN chooses one of three developmental actions – no action, adding a new voxel, and removing the voxel – by three neurons. The second of the two output neurons shown in Fig. 2 enables the ANN to determine one of four materials to add to the voxel. The value of input and output neurons is defined by Eqs. (1) and (2). We use the ReLU as an activation function for each neuron. All outputs are converted into zero or one depending on the output neuron of the ANN.

$$u_i = \sum_j \omega_{ji} v_i \tag{1}$$

$$v_i = \max(0, u_i) \tag{2}$$

If the addition rule is selected, the new voxel is created toward the direction determined by the ANN unless no empty space is retained, If not, the operation is aborted. All voxels are able to create a new voxel toward six kinds of direction from adjacent space every 100 steps of the development stage. Meanwhile, all voxels deal with the deletion rule as a priority even before the addition is activated, in which case the addition is not executed; however, no deletion happens when the total number of voxels equal to one or robot is separated into two or more parts. The total number of voxels is as many as 1331 to prevent a declining calculation speed if the number of created voxels exceeds each limitation. Figure 2 shows an overview of the addition and deletion processes (see right figure).

2.4   Age-Fitness Pareto Optimization

Age-fitness pareto optimization (AFP) is an evolutionary algorithm for multiple objective optimization proposed by Schmidt [28]. Adjusting weights in the ANN is required to obtain the desired dynamics because the initial statement of the network randomly outputs a value and, in many cases, these values are meaningless for achieving a task. It is difficult, however, to explicitly optimize a set of weights in the ANN when the virtual robot has a more complicated morphology, many actuators, and has been given a complex task or situation. We adopt AFP to improve the connection weights of our ANN because AFP showed high performance for many optimization benchmark problems and it is also applied to optimize multiple finesses. The AFP can treat a set of weights of the ANN as a real-valued vector and approximately improve these weights while maintaining the diversity of the vectors. These vectors are known as individuals in evolutionary algorithms. Basically, the AFP has concepts of a population, i.e., a set of individuals and a generation, i.e., the number of times of improvement. The classic evolutionary algorithm repeatedly conducts three evolutionary operations – crossover, mutation, and selection – to multiple individuals to retain the good features of the previous population in the next population. The crossover operation creates a new individual by exchanging a part of two parent individuals. The mutation operation creates a new individual by changing some elements of the parent individual. The selection operation chooses more appropriate individuals based on any evaluation criterion. Here, as the important factor of the

AFP, there is the concept of aging [29, 30]. All individuals have age and the age merely increases while the generation increases. The AFP optimizes the fitness function with the age and the solution is considered more optimal if the age is below that of other individuals. Optimizing the fitness function by minimizing the age prevents the early convergence of solution search. Thus, the AFP adds a rule of adding a zero-age individual into the current population in three evolutionary operations of the classic evolutionary algorithm.

Our crossover operation chooses a couple of individuals with the crossover probability Pc and selects one of the output units and by adopting BLX-α as the crossover. BLX-α determines new individual y_i from two parent individuals $x_1 = (x_{11}, x_{12}, \ldots)$ and $x_2 = (x_{21}, x_{22}, \ldots)$ based on Eqs. (3) and (4).

$$y_i = \alpha d_i r + x_{1i} \tag{3}$$

$$d_i = x_{2i} + x_{1i} \tag{4}$$

The basic concept of BLX-α is that a more optimal individual exists in the solution space between two parent individuals. Our simulation does not include the mutation operation because BLX-α includes the meaning of the mutation operation. Our selection operation also randomly chooses a couple of individuals with the more optimal individual overwriting another individual, and it is conducted by comparing the fitness function and age between two individuals. If the fitness value of one individual is larger than that of another individual and the age is below that of another individual, the inferior individual is removed from the current population. The selection is named tournament selection and all individuals are retained unless they are removed from the population.

2.5   Dynamics Computation

As mentioned above, our simulation uses Voxelyze to simulate a soft-bodied robot and fluid motions. Our simulation model calculates buoyancy F_b and drag F_d to represent resistance in fluid. Basic translational and rotational motion at the center of the mass are used to describe the motion of one voxel. Eqs. (5) and (6) show the equation of motion,

$$F = m\frac{dv(t)}{dt} \tag{5}$$

$$T = \frac{dL(t)}{dt} \tag{6}$$

where F is the force vector, m the mass of the voxel, v the linear velocity of the voxel, t the time, T the torque, and L the angular velocity of the voxel. The conceptual design in Fig. 3 is intended to clarify our approach. We employ thermal expansion for the muscle voxels. At the end of the development stage, the muscle voxels vibrate depending on the frequency ω, the amplitude A, and phase shift ϕ, and those two values are used in Eq. (7).

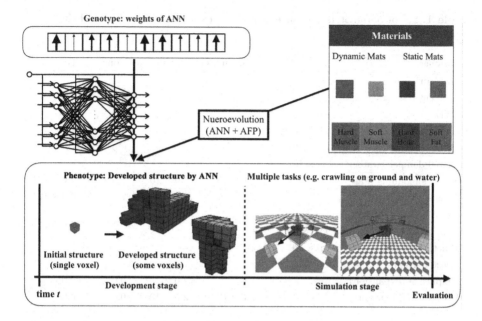

Fig. 3. Concept of our evolutionary strategy for robot development. Our simulation has two simulation stages: development and locomotion. The genotype of our simulation means a set of weights in the ANN and the phenotype is a morphology developed by the ANN from a single voxel during the development stage. The robot is evolved to accomplish a crawling task on the ground and underwater.

$$K_t = A\sin(2\pi\omega t + \phi) \tag{7}$$

where K is the temperature in the voxel, t the current simulation time, ϕ is the minimum angle between a normal vector of six surfaces of the voxel and the normalized vector from the voxel centroid to the target source. If all the surfaces of voxels have contact with other voxels or if the inner product of those vectors is smaller than zero or there is no target source, the angle ϕ equals zero. Voxelyze cannot calculate fluid forces in the default situation. Implementing calculation expressions supports buoyancy Fb and drag Fd in fluid [31], defined in Eqs. (8) and (9),

$$F_b = \rho V g \tag{8}$$

$$F_d = \frac{1}{2}\rho S C_d u_f^2 \tag{9}$$

where ρ is the fluid density, V the volume of the voxel, and g is the gravitational acceleration. Further, S is the voxel surface area of each direction, C_D is the coefficient of drag force, and u_f the relational velocity between the voxel and the fluid.

Table 1. Set of parameters for our evolutionary experiment with physical simulation by Voxelyze

Parameter	All	Hard muscle	Soft muscle	Bone	Fat
Size [cm^3]	$1.0 \times 1.0 \times 1.0$	-	-	-	-
Ambient temperature [C]	30.0	-	-	-	-
CTE	2.0×10^{-2}	-	-	-	-
Collision damper	5.0×10^{-1}	-	-	-	-
Global damper (ground)	1.8×10^{-5}	-	-	-	-
Global damper (water)	8.9×10^{-4}	-	-	-	-
Young's modulus [Pa]	-	1.0×10^7	1.0×10^6	1.0×10^7	1.0×10^6
Density (voxel)	-	1.1×10^3	1.1×10^3	2.0×10^3	9.0×10^2
Density (air)	1.2	-	-	-	-
Density (water)	9.95×10^3	-	-	-	-
Kinetic friction	5.0×10^{-1}	-	-	-	-
Static friction	6.0×10^{-1}	-	-	-	-
Amplitude (expansion)	-	7.0	7.0	0.0	0.0
Period [s] (expansion)	-	2.0×10^{-2}	2.0×10^{-2}	0.0	0.0

3 Experiments

3.1 Experimental Details

The evolved robots were analyzed in different environments from the viewpoint of morphological evolution and behavioral control. Our experiment prepared ground and underwater environments in virtual space and provided a pushing task involving a box object for the robots developed by the ANN. The evolution of the robot is compared in the different environments by dividing the experiment into three parts, which are (1) evolution on the ground, (2) evolution underwater and (3) multiple objective evolutions in both environments. Figure 3 shows the evolutionary concept of the NE and the physical simulation by dynamics computation. The development stage of the simulation is for 100 time steps and the morphology of the robot is updated every 1 step (=0.01 [s]) by the calculation of the ANN. After the end of the development stage, the simulation transits the locomotion stage. For the locomotion stage, the developed robot is simulated for 10.0 [s] in each environment and the behavior is evaluated at the end of the locomotion stage. In the case of multiple objective evolutions in both environments, the locomotion stage sequentially executes two simulations in both environments, after which we evaluate each behavior. The side length of single voxel is 1.0 cm. The box object is built by using $3 \times 3 \times 3$ voxels, and it is placed along the x-axis 15 voxels away from the center of the design space, which is $11 \times 11 \times 11$ voxels. The experiment consists of 30 runs, each with a population size of 30, evolved for 200 generations. Tables 1 and 2 present a set of parameters that were used in the evolutionary experiments.

Table 2. Set of parameters for the evolving artificial neural network and the age-fitness pareto optimization

Evolving artificial neural network		Age-fitness Pareto optimization	
Parameter	Value	Parameter	Value
Layer	3	Runs	30
Input	6	Population size	30
Hidden	20	Generations	200
Output	5	Tournament size	2
Bias	1	Crossover rate	0.9
Weight range	$[-1.0{:}1.0]$	Crossover α	0.5

3.2 Tasks and Penalties

The aim of the task is to determine how the robot pushes the box a long distance, and the robot needs a way to be able to crawl to the box and the body structure that moves while the robot is pushing the box. Thus, the task achievement detects whether to shift the box from the initial position or not. The behavior of the robot when carrying out the task evaluates the fitness function f (Eq. (10)).

$$f = \frac{1.0 + D_{box}}{1.0 + D_{robot}} \tag{10}$$

where D_{box} is the distance between the initial position of the box and the final position of the box, and D_{robot} is the distance between the final position of the robot and that of the box. The fitness function means maximizing the distance the box moves and minimizing the distance between the box and the robot. If the robot cannot achieve moving the box, the value of the distance the box moves equals zero and the situation inhibits the evolution of the robot. In order to prevent the situation, the minimum value of the evaluation for the movement of the box is defined as 1.0. Besides, the minimum value of the evaluation for the distance between the box and the robot equals 1.0 to prevent the distance from becoming zero by using the equation. Therefore, the minimum value of the fitness function becomes a positive value.

As the result of the calculation of the ANN, if the body of the robot separates into two parts or more, the individual AFP is replaced by a new individual with random values and repeatedly processes the development stage until the robot develops a single body. In case all voxels of the robot are static material voxels, the individual is also replaced by a new random individual. This is the reason why the evolution of the AFP is not able to gradually create the actuated robot when there are many static robots in the population.

4   Results

Videos of our soft-bodied robots locomotion are available at http://www.
junogawa.com/evolutionary-soft-robotics/.

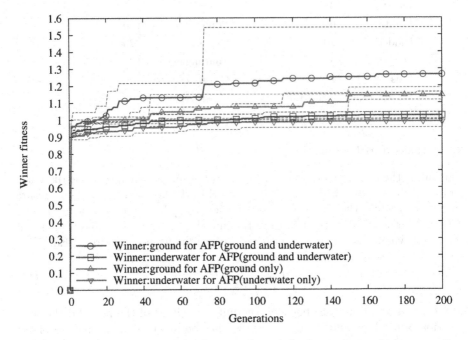

Fig. 4. Time series changes in the fitness value of the best robot, which means the
winner in the current population, in each experimental environment. These results are
the average values of 30 runs by the single and multiple objectives AFP. If the fitness
value exceeds 1.0, the robot surely has contact with the target box object.

4.1   Single vs. Multiple Objectives Optimization

In order to quantitatively evaluate the performance for single and multiple objective optimization, Fig. 4 shows the time series changes of the best robot in each
experimental environment, which are on the ground only, underwater only, and
both of these environments. In Fig. 4, the best robot by the multiple objectives
AFP is higher than the single objective AFP in both environments. It is clarified
that the existence of the ground surface and the difference between both environments contributed some specific effects to the behavior of the robot because
there is a difference between the time series changes of the evaluations on the
ground and underwater. The multiple AFP always retains the best robots in
both environments in the current population, and it mechanically composes the
structure of the best robots in both environments in the process of evolution.

The result shows that the multiple objectives AFP discovers the robot morphology with better control systems earlier than the evolved robot in a single environment.

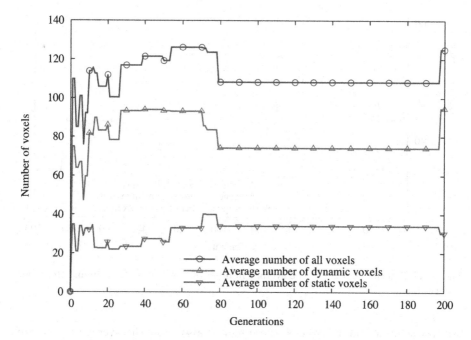

Fig. 5. Time series changes in the average number of voxels used for evolving the robot on the ground.

The search for the best robot underwater converges earlier than the search on the ground. The task difficulty and the replacement between the separated robot and a new robot would cause the convergence. The problem of early convergence is resolved by adjusting the task difficulty; for example, changing the simulation time in each environment or the initial position of the box. The destruction of the separation robot and the creation of a new robot are introduced into the AFP optimization to avoid the separation of the body of the robot. As a result of using the replacement operation, the evolutionary search increasing the elements of the random search; however, it is easy to discover better robots than with a normal AFP in each generation.

4.2   Robot Size and Material Types

Our experiments calculate the number of voxels in the robot to analyze the relationship between the size of the robot and the task accomplishment. Figure 5 shows the average number of voxels in the best robot on the ground, and Fig. 6 represents the time series changes in the average number of voxels in the

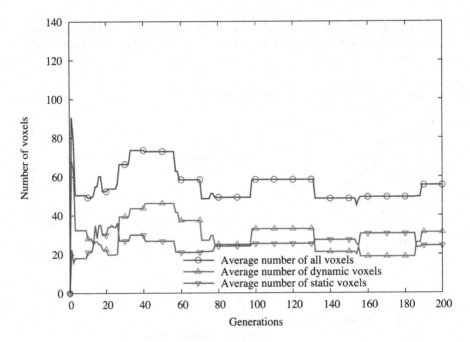

Fig. 6. Time series changes in the average number of voxels used for evolving the robot underwater.

best robot underwater. Besides, these figures also show the average number of dynamic voxels (hard muscle and soft muscle) and the average number of static voxels (bone and fat) in the robot. In the case of our evolutionary experiment on the ground, the total number of voxels in the best robot is approximately 110, and the number of dynamic and static voxels is approximately 75 and 35, respectively. The best robot on the ground accounts for 8% of the design space. In the evolution underwater, the number of used voxels in the best robot is about 60, and this value is similar to half of the number of voxels in the robot on the ground. Figure 7 shows the rate of dynamic and static voxels in the best robot in each environment. There is no large difference between the number of static voxels in both environments from Fig. 7. Therefore, our evolutionary approach acquired a robot structure that includes many muscle voxels, which directly produce power from the friction on the ground surface, in the evolution on the ground. As a result of the evolution underwater, the size of the robot is smaller than that of the robot that evolved on the ground. This robot is able to produce the power for accomplishing the same task underwater. Thus, it was clarified that the importance of muscle voxels is less important than the behavior on the ground when the structure of the robot is developed to optimize the dynamics among the vibration power by muscle voxels and the drag forces of all voxels from the water for the task.

4.3   Morphologies

In order to understand the morphological change in the best robot for each task we visualize the time series of the evolved structure of the best robot in Figs. 8 and 9. The morphology of the evolved structures is actually quite indescribable looking during early generation in both environments. During the end of evolution on the ground, these structures gradually transit to morphologies such as a wing (see first line in Fig. 8), a slug (see second line in Fig. 8), a dome (see fourth line in Fig. 8), and some limb-like parts. Partly, the appearance of the robot with four limb-like parts (see fifth line in Fig. 8) resembles that of a real robot or real four-legged insects. Basically, the evolved robots have parts to catch or push the box and the robots use those parts to retain the box near those parts until the simulation finishes accomplishing the task. Most of the robots underwater gradually become very simple structures such as fish or a propeller for the evolution. For the left robot of the third line and the middle robot of the fifth line In Fig. 8 the robot left of the third line and the robot in the middle of the fifth line are those that are the best in both environments at generation. Our evolutionary approach was able to discover that the common structures have a morphological adaptation to crawl in both environments in the evolution process.

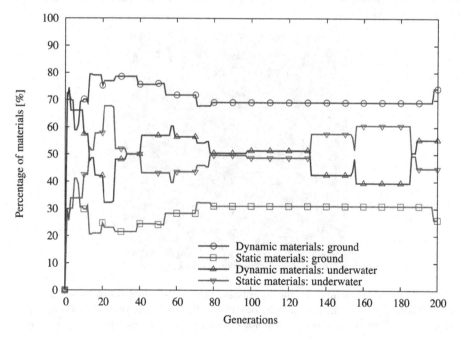

Fig. 7. Rate of used dynamic voxels in the best robot in each evolutionary experiment. The percentage of dynamics voxels in the robot on the ground is about 70%. This result means that the robot crawling on the ground needs many actuators in the body. Underwater, the difference between these rates is insignificant, and it shows that both material voxels have the role of obtaining power at the same level.

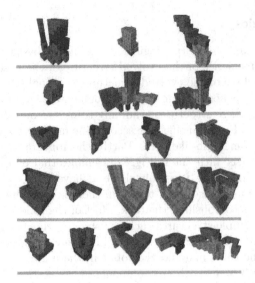

Fig. 8. Time series changes in the morphologies of soft-bodied robot to crawl on the ground environment. From left to right, the morphology of the best robot is transited. From top to bottom, we show the evolved morphology for five examples for the result of runs of the AFP.

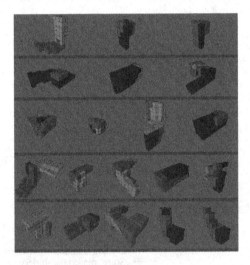

Fig. 9. Time series changes in the morphologies of soft-bodied robot to swim underwater. From left to right, the morphology of the best robot is transited. From top to bottom, we show the evolved morphology for five examples for the result of runs of AFP.

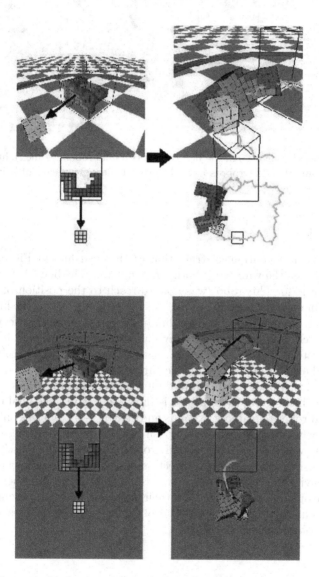

Fig. 10. Trajectory of behavior of the robot in the middle of the fifth line in Fig. 9. This robot is able to reach the box object in both environments.

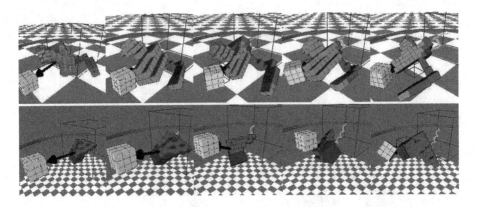

Fig. 11. Time series changes in the behavior of two final robots to the right of the fifth line in Figs. 8 and 9. Those robots have the evolutionary experience of being the same structure before generation.

4.4   Behavior

According to the evolutionary transition of the fifth line in Fig. 9, the same robot is chosen as the winner for both environments. The behavior of this robot is shown in Fig. 10. This robot was able to reach to the position of the box in both environments. As a result, in Fig. 10, the ground robot is pushing the box while walking around by using four limb-like parts; however, the underwater robot was reached by twisting its body and swimming as though it is drawing an arch trajectory. This result showed that the same soft-bodied robot changes its adaptive behavior depending on the surrounding environment.

After that, the robot finally evolved into different morphologies in each environment. The behavior of those robots is shown in Fig. 11. The final morphology of the ground robot repeatedly expands and contracts by two legs on a diagonal, and the robot was able to effectively push the box by moving these legs. Then the final underwater robot was gradually rotating by using the muscle voxels like the tail of a fish to advance in the direction of the box. The robot also acquired pushing behavior. Moreover, both robots retained their pushing behavior when the position of box was changed by their behavior (Fig. 12). From the viewpoint of morphology, the ground robot evolved like a terrestrial creature and the underwater robot evolved like a fish.

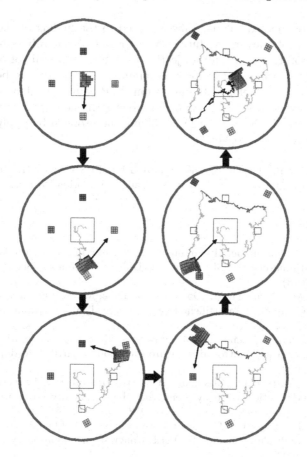

Fig. 12. Simulation prepares the evolved robot for pushing the yellow box and four boxes. We positioned them on the ground. The robot changes the target box depending on the current elapsed time. First is the yellow box and the simulation time is for 10.0 s, after which the robot changes the target box counterclockwise every 20.0 s. The simulation result confirmed that the robot was able to crawl in any direction. (Color figure online)

5   Conclusion

In this work we proposed a novel way of developing evolutionary soft robotics from the viewpoint of morphological evolution with adaptation in different environments. Our method was able to acquire more interesting morphologies of a soft-bodied robot by reducing the human bias in the morphological evolution. By introducing the role of sensing function into a voxel, our evolved robot was able to conduct more controllable and adaptive behavior than the evolved rigid-bodied robots in most conventional evolutionary robotics approaches. We ascertain that there exist many unintentional and complex robots depending on the different experimental situation. Besides, our simulation would be available for creating more robust soft-bodied robots. This work led to the finding of the possibility of evolutionary robotics in the future development of soft robots toward environmental adaptation and multi-objective learning. The result of optimization continues to remain incomplete for maintaining the diversity of robot morphology because the morphology obtained from AFP converges at an early stage. In future, we would need to focus on improving the acceleration of the simulation speed with the NE. We suggest solving the problem by applying a GPU acceleration mechanism to Voxelyze.

Acknowledgement. This work was supported by JSPS KAKENHI Grant Number 17K12756 and the University of Aizu Robot Valley Promotion Project.

References

1. Furukawa, M., Watanabe, M., Fukumoto, A., Suzuki, I., Yamamoto, M.: Swimming animats with musculoskeltal structure. IADIS Int. J. Comput. Sci. Inf. Syst. **7**(2), 152–164 (2013)
2. Furukawa, M., Morinaga, M., Ooe, R., Watanabe, M., Suzuki, I., Yamamoto, M.: "Behavior Composed" for artificial flying creature. J. Adv. Comput. Intell. Inform. **5**(7), 838–845 (2011)
3. Ooe, R., Suzuki, I., Yamamoto, M., Furukawa, M.: Study on evolution of the artificial flying creature controlled by neuro-evolution. J. Artif. Life Robot. **17**(3–4), 470–475 (2013)
4. Ohkura, K., Yasuda, T., Matsumura, Y.: Extracting functional subgroups from an evolutionary robotic swarm by identifying the community structure. In: Nature and Biologically Inspired Computing (NaBIC), pp. 112–117 (2012)
5. Auerbach, J., Bongard, J. C.: How robot morphology and training order affect the learning of multiple behaviors. In: IEEE Congress on Evolutionary Computation, pp. 39–46 (2009)
6. Auerbach, J.E., Bongard, J.C.: Evolving CPPNs to grow three-dimensional physical structures. In: Proceedings of the Genetic and Evolutionary Computation Conference, pp. 627–634 (2010)
7. Auerbach, J.E., Bongard, J.C.: On the relationship between environmental and morphological complexity in evolved robots. In: Proceedings of the Genetic and Evolutionary Computation Conference, pp. 521–528 (2012)
8. Lipson, H., Pollack, J.B.: Automatic design and manufacture of robotic lifeforms. Nature **406**(6799), 974–978 (2000)
9. Lessin, D., Fussell, D., Miikkulainen, R.: Adapting morphology to multiple tasks in evolved virtual creatures. In: The Fourteenth Conference on the Synthesis and Simulation of Living Systems, vol. 14, pp. 247–254 (2014)
10. Sayama, H.: Swarm chemistry. Artif. Life **15**(1), 105–114 (2009)
11. Sayama, H.: Morphologies of self-organizing swarms in 3D swarm chemistry. In: Proceedings of the 14th Annual Conference on Genetic and Evolutionary Computation, pp. 577–584 (2012)
12. Sayama, H., Wong, C.: Quantifying evolutionary dynamics of swarm chemistry. In: Advances in Artificial Life, ECAL 2011: Proceedings of the Eleventh European Conference on Artificial Life, pp. 729–730 (2011)
13. Pfeifer, R., Bongard, J.C.: How the Body Shapes the Way We Think: A New View of Intelligence. MIT Press, Cambridge (2006)

14. Sims, K.: Evolving virtual creatures. In: Proceedings of the 21st Annual Conference on Computer Graphics and Interactive Techniques, pp. 15–22 (1994)
15. Doursat, R., Sanchez, C.: Growing fine-grained multicellular robots. Soft Robot. 1(2), 110–121 (2014)
16. Doursat, R., Sayama, H., Michel, O.: A review of morphogenetic engineering. Nat. Comput. 12(4), 517–535 (2013)
17. Joachimczak, M., Suzuki, R., Arita, T.: Fine grained artificial development for body-controller co-evolution of soft-bodied animats. In: The Fourteenth Conference on the Synthesis and Simulation of Living Systems, vol. 14, pp. 239–246 (2013)
18. Joachimczak, M., Suzuki, R., Arita, T.: From tadpole to frog: artificial metamorphosis as a method of evolving self-reconfiguring robots. In: Proceedings of the Thirteenth European Conference on the Synthesis and Simulation of Living Systems (ECAL 2015), pp. 51–58 (2015)
19. Joachimczak, M., Wrobel, B.: Co-evolution of morphology and control of soft-bodied multicellular animats. In: Proceedings of the 14th Annual Conference on Genetic and Evolutionary Computation, pp. 561–568 (2012)
20. Geijtenbeek, T., van de Panne, M., van der Stappen, A.F.: Flexible muscle-based locomotion for bipedal creatures. ACM Trans. Graph. (TOG) 32(6), 206 (2013)
21. Cheney, N., MacCurdy, R., Clune, J., Lipson, H.: Unshackling evolution: evolving soft robots with multiple materials and a powerful generative encoding. ACM SIGEVOlution 7(1), 11–23 (2014)
22. Cheney, N., Clune, J., Lipson, H.: Evolved electrophysiological soft robots. ALIFE 14, 222–229 (2014)
23. Vohradsky, J.: Neural network model of gene expression. FASEB J. 15(3), 846–854 (2001)
24. Hiller, J., Lipson, H.: Multi-material topological optimization of structures and mechanisms. In: Proceedings of the Genetic and Evolutionary Computation Conference, pp. 1521–1528 (2009)
25. Hiller, J., Lipson, H.: Evolving amorphous robots. In: Artificial Life XII, pp. 717–724 (2010)
26. Hiller, J., Lipson, H.: Automatic design and manufacture of soft robots. IEEE Trans. Robot. 28(2), 457–466 (2012)
27. Hiller, J., Lipson, H.: Dynamic simulation of soft multimaterial 3D-printed objects. Soft Robot. 1(1), 88–101 (2014)
28. Schmidt, M., Lipson, H.: Age-fitness pareto optimization. In: Riolo, R., McConaghy, T., Vladislavleva, E. (eds.) Genetic Programming Theory and Practice VIII, pp. 129–146. Springer, New York (2011). https://doi.org/10.1007/978-1-4419-7747-2_8
29. Hornby, G.S.: ALPS: the age-layered population structure for reducing the problem of premature convergence. In: Proceedings of the 8th Annual Conference on Genetic and Evolutionary Computation, Seattle, pp. 815–822 (2006)
30. Hornby, G.S.: A steady-state version of the age-layered population structure EA. In: Riolo, R., O'Reilly, U.M., McConaghy, T. (eds.) Genetic Programming Theory and Practice VII, pp. 87–102. Springer, Boston (2009). https://doi.org/10.1007/978-1-4419-1626-6_6
31. Tolley, M.T., Kalontarov, M., Neubert, J., Erickson, D., Lipson, H.: Stochastic modular robotic systems: a study of fluidic assembly strategies. IEEE Trans. Robot. 26(3), 518–530 (2010)

Field Coverage for Weed Mapping: Toward Experiments with a UAV Swarm

Dario Albani[1,2(✉)], Tiziano Manoni[2], Arikhan Arik[2], Daniele Nardi[2], and Vito Trianni[1]

[1] ISTC, National Research Council of Italy, Rome, Italy
{dario.albani,vito.trianni}@istc.cnr.it
[2] DIAG, Sapienza University of Rome, Rome, Italy
{albani,manoni,arik,nardi}@diag.uniroma1.it

Abstract. Precision agriculture represents a very promising domain for swarm robotics, as it deals with expansive fields and tasks that can be parallelised and executed with a collaborative approach. Weed monitoring and mapping is one such problem, and solutions have been proposed that exploit swarms of unmanned aerial vehicles (UAVs). With this paper, we move one step forward towards the deployment of UAV swarms in the field. We present the implementation of a collective behaviour for weed monitoring and mapping, which takes into account all the processes to be run onboard, including machine vision and collision avoidance. We present simulation results to evaluate the efficiency of the proposed system once that such processes are considered, and we also run hardware-in-the-loop simulations which provide a precise profiling of all the system components, a necessary step before final deployment in the field.

1 Introduction

Swarm robotics can have a large impact in many application domains, especially when the tasks to be accomplished are distributed widely in space, and when there is room for collaboration among robots [5]. In such conditions, not only it is possible to profit of the parallel execution of tasks by multiple robots, but the execution of a single task is made more efficient thanks to collaboration, opening to the exploitation of solutions designed for multi-robot task allocation problems [7,11]. An application domain presenting expansive fields and needs for collaboration is certainly precision agriculture, whereby robotics technologies promise a remarkable impact [9]. Current practice is however far from exploiting the full potential of autonomous robots and multi-robot collaboration. Off-the-shelf remote sensing technologies with unmanned aerial vehicles (UAVs), for instance, rarely go beyond passive data collection with predefined mission plans [8,13]. Therefore, even in field mapping applications, much improvement is expected by the introduction of adaptive mission planning and collaboration among multiple UAVs [2,14].

A. Compagnoni et al. (Eds.): BICT 2019, LNICST 289, pp. 132–146, 2019.
https://doi.org/10.1007/978-3-030-24202-2_10

Previous work proposed swarm robotics approaches for UAVs engaging in a weed monitoring and mapping problem [1;2]. In this problem, weeds must be recognised from crops in order to create a precise weed-density map to be exploited for weed control operations (e.g., for variable-rate herbicide applications). Swarm robotics solutions are useful to parallelise the monitoring and mapping operations, and to have robust controllers scalable to different field sizes [2]. Additionally, the weed distribution across an expansive field is often heterogeneous, with weed patches infesting certain areas while other areas remain devoid of weed. In such conditions, it is useful to adaptively monitor only those areas where weeds are present—and to collaborate for that purpose—resorting to non-uniform coverage and mapping strategies [1,15]. These studies just provide a proof of concept in abstract simulations, and several relevant features that pertain to the real-world implementation have been overlooked. For instance, collision avoidance among UAVs was not taken into account, and the onboard weed detection through machine vision was streamlined by a simple mathematical model of the weed detection error [2].

In this paper, we move toward real-world testing with the following contributions. (i) We provide the full implementation of a weed coverage and mapping system for a UAV swarm, with a design tailored to the Avular Curiosity platform [3]. For experimental purposes, we set up an experimental indoor arena whit a green carpet and pink golf balls to represent weeds, following the rules of the 2017 Field Robot Event [6], a renowned competition for autonomous farming robots. This choice drastically simplifies the vision routines, but ensures that all the components of the control loop are taken into account. (ii) We provide an improved implementation of the stochastic coverage and mapping presented in [2], including collision avoidance among UAVs and onboard vision. We test the performance of the deployed algorithm in simulation for a wide range of parameterisations, identifying the most suitable parameters sets for efficient coverage and mapping. (iii) We integrate the hardware platform into the simulation environment and we perform hardware-in-the-loop (HIL) simulations to profile the different components of the onboard system. We propose HIL simulations as a convenient way of testing swarm robotics controllers before actual deployment, especially when physical interactions are not needed. (iv) We provide proof-of concept videos of the real flying system.

The paper is structured as follows. Section 2 describes the design of the different components implementing the stochastic monitoring and mapping strategy. Section 3 discusses the results of simulations for field coverage, weed mapping and HIL simulations. Section 4 concludes the paper discussing the results obtained and the future steps necessary for field deployment of UAV swarms.

2   Experimental Setup

Weed monitoring and mapping requires UAVs to fully cover a field, detect the presence of weeds and map their exact position. We considered a mockup version of the real problem by simplifying the vision requirements, so that we can focus on the development and testing of the swarm-level strategy. In this mockup version, weeds are represented by pink golf balls placed on a green carpet, making their detection a relatively simple task. Here, our main objective is to develop and test a simple and reliable coverage and mapping algorithm in realistic conditions, and to flexibly switch between simulation and execution on the UAV platform, also mixing the two with HIL simulations.

2.1   Hardware

The Avular Curiosity platform is a small square quad-rotor of about 40 cm side and 1 kg weight (see Fig. 1) [3]. The platform provides dedicated hardware for low-level flight control and high-level mission execution. The low-level control unit consists of a PX4 micro-controller that acts as a bridge between the high level abstraction and another PX4 that implements the autopilot. The low-level unit integrates a variety of sensors for motion control, such as a real-time kinematic global navigation satellite system unit (RTK-GNSS), a dual 9-DOF IMU, a barometer and a laser range-finder for altitude control, and the Ultra-Wide Band (UWB) system for indoor self-localisation. These sensors support precise absolute localisation both indoor and outdoor, hence allowing waypoint navigation and simple trajectory generation.

The high-level control unit consists of a Raspberry Pi3 (RPi), a small and affordable programmable device characterised by low energy consumption. The role of the RPi is to support the overall mission execution strategy (see Sect. 2.2), which includes both onboard vision and communication with other UAVs within the swarm. The latter is achieved thanks to the Digi ZigBee, a specific module coming from the Xbee family of communication devices.

2.2   Software

The proposed implementation of the high-level mission execution strategy exploits the Robot Operating System (ROS) structure, in particular ROS Kinetic running on the RPi platform, and is illustrated in Fig. 1. The overall framework improves over the stochastic coverage and mapping strategies presented in [2] by refining the reinforced random walk model including collision avoidance (see also Sect. 2.3), by removing inefficient computations and by adding onboard vision (Sect. 2.4).

More in detail, the proposed implementation offers three main ROS packages: (i) `Core`, (ii) `Perception` and (iii) `Communication`, each one responsible of a specific task. The `Core` package is responsible for defining the navigation of the UAV between different areas of the field, assuming knowledge of the absolute position of the drone, either from GNSS outdoor or from the UWB indoor. The `Perception` package is responsible for image acquisition and processing. It presents two different nodes, one for communication with an external camera (i.e., the Raspicam connected to the RPi) to be used for real field tests, and another one used in the HIL simulations that loads real images from the local dataset and processes them on-line. The last package is `Communication`, which enables swarm operations and allows drones to perform cooperative field mapping as well as communication between high-level and low-level control units.

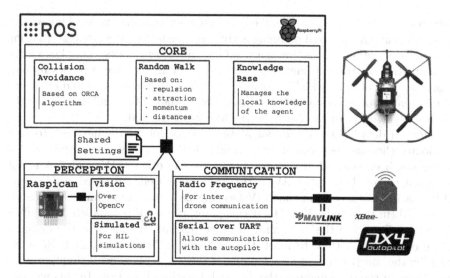

Fig. 1. Proposed implementation of the mission execution strategy for the Avular Curiosity platform [3], which is shown in the top right side. The development is made on the Raspberry Pi3 platform using ROS running over a custom version of Ubuntu Mate 16.04. There are three main ROS packages, namely (i) `Core` (ii) `Perception`, and (iii) `Communication`. Each package allows to run any of different nodes it embeds.

The node responsible for the latter is the `Serial` node, which enables direct communication with the low level autopilot through the RPi serial port. Communication with other drones is made possible by the `Radio Frequency` node that forwards and processes information to and from the ZigBee external module, implementing also the selected information-aware broadcasting protocol [2]. Information sharing is made possible by the Micro Air Vehicle Link (MAVLink) communication protocol, responsible of encoding and decoding both default and custom messages.

2.3   Stochastic Coverage and Mapping

The basic navigation strategy implemented by the `Core` package derives from previous work [2], and improves it in several ways. As in previous work, we consider a field divided in square cells of 1 m side. The UAVs have access to their global position (either through GNSS or UWB) and can communicate reliably with each other.[1] We assume a maximum density for the point of interests in each cell. For instance, in the indoor tests here presented we assume that each cell contains up to 12 pink golf balls. To accomplish the coverage task, each cell needs to be visited at least once and by at least one UAV while, to accomplish the mapping task the number of balls within the cell should be reliably

---

[1] We ignore here communication limitations, which have been suitably accounted for through information re-broadcasting protocols [1,2].

detected. Similarly to [2], the `Knowledge Base` node contains the local knowledge of the current coverage and mapping activities, which also integrates information received from other UAVs through the `Communication` nodes. This knowledge is completely distributed, meaning that each UAV in the swarm posses its own representation of the world. The local knowledge is constantly updated by means of information observed locally or received from other UAVs, and determines the navigation strategy implemented in the `Random Walk` node.

More in details, at every decision step, a UAV selects the next cell to visit randomly choosing one cell from a valid set $\mathcal{V}$. A cell is considered valid if it has not been covered or mapped before by any other UAV, and if it is not occupied/targeted by other UAVs within the swarm. Cells are added to $\mathcal{V}$ in order of increasing distance from the UAV (see Fig. 2a), until a maximum number V is reached (in this study, $V = 1$). At this point, the set $\mathcal{V}$ is completed including all valid cells within the maximum distance reached. In this way, the distance to be covered by a UAV to reach the next cell is always minimised. Here, we simplify the construction of $\mathcal{V}$ by looking at the whole plane, while in [2] priority was given to the semi-plane in the forward direction of motion. In this way, we save precious computational time for implementing the strategy on the RPi.

We consider here a reinforced random walk [17], in which a directional bias is given by three components, as shown in Fig. 2a: (i) the individual momentum m_h, a unit vector in the direction of motion of the UAV h resulting in a correlated random walk; (ii) the repulsion vector r_h from all other UAVs in the swarm; (iii) the attraction vector a_h towards cells previously marked as attractive by other UAVs with virtual beacons $b \in \mathcal{B}$. Attraction and repulsion vectors are computed as follows:

$$r_h = \sum_{u \neq h} S(\boldsymbol{x}_h - \boldsymbol{x}_u, \sigma_a), \ a_h = \sum_{b \in \mathcal{B}} S(\boldsymbol{x}_b - \boldsymbol{x}_h, \sigma_b), \ S(\boldsymbol{v}, \sigma) = 2e^{i\angle v}e^{-\frac{|v|}{2\sigma^2}}, \ (1)$$

where $\boldsymbol{x}_i$ represents the position of agent/beacon i, and $S(\boldsymbol{v}, \sigma)$ returns a vector in the direction of $\boldsymbol{v}$ with a Gaussian length with spread σ. With respect to [2], we simplify the way in which beacons are placed and removed. Here, an agent places a beacon on a cell if something has been detected, and only if no other beacon is present. Beacons are removed from cells that are reliably mapped.

The selection of the next cell to visit is performed randomly using the vector $\boldsymbol{v}_h = m_h + r_h + a_h$ as a bias. Each cell $c \in \mathcal{V}$ is assigned a probability $P_c = u_c / \sum_{i \in \mathcal{V}} u_i$ of being selected, where the utility u_c is computed according to the angular difference $\theta_c = \angle(\boldsymbol{x}_c - \boldsymbol{v}_h)$ as follows:

$$u_c = C(\theta_c, 1 - e^{\frac{|v_h|}{2}}), \quad C(\theta, p) = \frac{1}{2\pi}\frac{1 - p^2}{1 + p^2 - 2p\cos\theta} \quad (2)$$

where $C(\cdot)$ is the wrapped Cauchy density function with persistence p. Differently from [2], we use the length of the bias vector $\boldsymbol{v}_h$—filtered by a smooth exponential ceiling function—to modulate the persistence of the random motion, so that a small module results in lower directional bias in the cell choice.

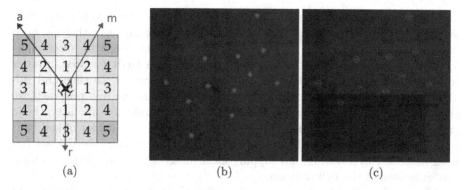

(a)   (b)   (c)

Fig. 2. (a) The RRW gives priority to neighbouring cells—numbered for increasing distance from the UAV placed in the center—and to cells in the direction of the resultant vector from the individual momentum m, the repulsion from other drones r and the attraction to beacons a. (b–c) Weeds are represented by pink golf balls over a green carpet. Circles around the balls represent the output of the vision module. Artificial disturbances are included in the left image to obtain realistic output from the visual processing. Disturbances represent illumination variance, shadows and motion blur. (Color figure online)

One of the biggest limitations of [2] is the absence of collision avoidance, which is required for real world deployment. Here, the navigation strategy includes the `Collision Avoidance` ROS node, which is responsible to generate sage and accurate trajectories. To this end, information about the position of other UAVs (as broadcasted through the `Communication` node) is made available by the `Knowledge Base` node. Based on such information, we implement the Optimal Reciprocal Collision Avoidance (ORCA) method [4,18]. ORCA is based on the knowledge of the position and velocities of the agents that may interfere with the planned trajectory. To avoid collisions, ORCA assigns part of the responsibility of implementing a correct manoeuvre to all agents involved. The method calculates all possible collisions within a time interval τ (in this work, $\tau = 1\,\mathrm{s}$). To this end, an agent is approximated by a disk with a safety radius r_o, which indicates the area around the agent that should not be violated. Then, all possible collision-free velocities are computed and the one that remains closer to the original velocity is selected. In this work, we integrate ORCA with the reinforced random walk strategy by letting the former change the trajectory to reach the desired destination. In cluttered conditions, collision avoidance may be very time-consuming. To avoid deadlocks, we let UAVs abandon their current target in favor of a new one if the flying time exceeds twice the expected value.

2.4 Onboard Vision Module

The goal of the `Vision` package is to implement the routines necessary to detect and count pink golf balls over a green carpet. Field experiments will add ROS nodes for the real weed recognition systems based on previous and ongoing studies [10,12]. To detect pink golf balls, a simple blob-detection algorithm has been implemented using OpenCV. We used a trivial procedure consisting of circularity filtering and converted the image color space from BGR to HSV in order to make the extraction of the pink golf balls more accurate. To tune and validate the blob detection algorithm, as well as for HIL simulation purposes, we generated a large dataset by placing pink golf balls randomly over a synthetic grass carpet within a $1\,\mathrm{m}^2$ area (see Fig. 2b). We varied the number of golf balls from 0 to 12, and for each size we collected 20 images. Finally, the dataset has been extended by rotating and flipping each image generating further unique configurations of the ball positions.

After tuning, the developed blob detection algorithm perfectly detects and counts the balls within the dataset. However, images taken from the UAVs while flying will never be as sharp and bright as the ones in the collected dataset. To evaluate the effectiveness of the devised mapping strategy under realistic working conditions, we artificially add disturbances to the images, to represent (i) changes in global illumination, (ii) shadows locally affecting portions of the image, and (iii) motion blur due to sudden rapid movements of the UAV. We simulate these disturbances by reducing the contrast of the image globally as well as within randomly generated blobs in the image, and then we apply a convolution with an averaging Gaussian kernel with a random direction (see Fig. 2c). To evaluate the performance of the Vision, we traversed all of the dataset 100 times and for each image we added artificial disturbances. We tuned the artificial disturbances to obtain an overall 22% error (mean: 0.2175 standard deviation: 0.018) so to ensure realistic and competitive tests. Indeed, given such error, it is not possible for the UAVs to know whether a cell is reliably mapped or not. In this paper we approximated this decision by marking a cell as reliably mapped within the Knowledge Base when the number of detected balls does not differ from a previous passage. In this way, at least two passages over the same cell are necessary to mark it as reliably mapped. The mapped state is then broadcasted to all other UAVs in the swarm via the Communication package.

3   Results

As mentioned above, the software we developed can be seamlessly integrated with simulations or executed by the UAV platform [3]. We present here simulation results to study both simple coverage by UAV swarms (Sect. 3.1) and weed mapping (Sect. 3.2). We also present results of HIL simulations, whereby the UAV platform is integrated within the simulator and used to execute—without flying—the coverage and mapping algorithm (see Sect. 3.3).

3.1   Field Coverage

The simple coverage problem requires that UAVs inspect every portion of the field by visiting (and inspecting) each cell at least once. Ideally, every cell

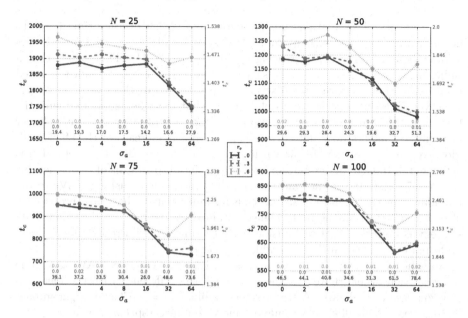

Fig. 3. Coverage time (absolute t_c and relative t_c^*) for varying swarm size N. Each panel shows the average results over 150 runs varying the Gaussian spread σ_a used to compute the repulsion vector r and the safety radius r_o that parametrises the ORCA algorithm. Errorbars represent the standard error. The number of collisions detected in each conditions is printed in the bottom of each panel, and color-coded according to the corresponding value of r_o.

is visited only once by one UAV and the coverage time t_c is minimised. In this study, we consider a field of $M = 50 \times 50 = 2500$ cells, and swarm size $N \in \{25, 50, 75, 100\}$. Following the approach taken in [2], we consider the absolute coverage time t_c as well as the time t_c^* relative to a lower bound computed as $t_1 M/N$, where t_1 is the time taken by a UAV to move between two adjacent cells. Focusing on coverage only, we disable the Vision node, and cells are marked as covered as soon as visited by a UAV, so that they are not visited a second time. Also, virtual beacons are not used for coverage. We instead focus on the interaction between the random walk, repulsion from other UAVs (as parametrised by the Gaussian spread $\sigma_a \in [0, 64]$), and collision avoidance (as parametrised by the ORCA radius $r_o = \{0, 0.3, 0.6\}$, considering collision avoidance disabled when $r_o = 0$). We also count the number of potential collision events detected anytime two UAVs get closer than 0.3 m. Note that these events do not affect the UAV motion in simulation (i.e., UAVs continue their mission even after a potential collision has been detected). In this way, we can evaluate the effects of collision avoidance on the overall performance.

The results of simulations are presented in Fig. 3. Coverage time tends to decrease with larger repulsion among UAVs, especially for $\sigma_a > 8$ until it hits a lower bound and starts increasing again (e.g. $\sigma_a > 64$). This means that a sufficiently high repulsion is necessary to ensure that UAVs remain separated from each other to cover different areas. As the size of the swarm increases, the coverage time decreases in absolute terms, but not when compared with the lower bound, corresponding to a sub-linear increase of efficiency (i.e. a linear increment of agents does not correspond to a linear increase of the efficiency). Moreover, as the size of the swarm increases we observe that high repulsion is detrimental. Indeed, high repulsion values associated to swarms of a considerable size (i.e. 75 and 100) do not aid the overall performance, pushing UAVs to the boundaries of the environment and causing inner cells to have low probability of being visited. Still, the performance obtained by the reinforced random walk strategy is remarkable, if we consider that the lower bound represents an ideal performance that is hardly achievable in practice.

In general collision avoidance results in slower coverage as UAVs take time to avoid each other. However, when $r_o = 0.3$, the difference with the ideal no-collision case is negligible. A larger safety radius ($r_o = 0.6$) has a larger impact because resolving collisions is generally more complex, especially when $\sigma_a > 16$. In these conditions, the strong repulsion makes UAVs move in a more directed way, hence interfering more with each other. Despite the slightly longer coverage time, ORCA successfully manages to avoid collisions among UAVs. The number of detected collision events is very high when ORCA is disabled, but practically all collisions are avoided when avoidance is enforced.

3.2   Weed Mapping

Once studied the effect of collision avoidance on the field coverage efficiency, we move to study the ability to precisely map the field for the presence of weeds, here represented by pink golf balls. We consider again fields of $M = 50 \times 50$ cells containing 5 patches of weeds, each represented by a square of 6×6 cells where the distribution of balls follows a 2D Gaussian, having (higher density in the center and lower density in the periphery. Each weed-infested cell has an associated image from the dataset, over which artificial disturbances can be added at runtime, as described in Sect. 2.4. We study both the conditions with and without such disturbances, to evaluate their effect on the mapping strategy. As mentioned in Sect. 2.4, a cell is marked as reliably mapped when the number of balls detected is equal to the number previously stored in the Knowledge Base. Hence, when no disturbance is added, two visits per cell are required even when no weed is present. More visits may be required in case of perception errors. UAVs perform a reinforced random walk under the influence of both repulsion from other agents ($\sigma_a \in \{0, 2, 4, 8, 16, 32\}$) and attraction to virtual beacons ($\sigma_b \in \{0, 2, 4, 8, 16, 32\}$). We fix in this case the value of the ORCA radius $r_o = 0.3$, which provides safe conditions for collision avoidance with negligible performance cost, as discussed in Sect. 3.1. To evaluate the performance of the system, we consider here the coverage time t_c and the mapping time t_m.

The latter corresponds to the time in which all weed-infested cells are marked as reliably mapped. Additionally, we look at the mapping accuracy by recording the average detection error with respect to ground truth.

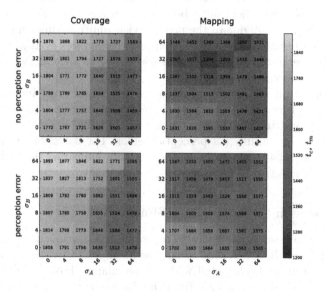

Fig. 4. Comparison of the coverage and mapping time for $N = 50$ robots. Each cell in the heatmap represents the average of 150 runs.

Figure 4 shows the average coverage and mapping time for $N = 50$ UAVs.[2] When no error in perception is considered (top-left and top-right panels), coverage time decreases with increasing repulsion among the UAVs—similarly to what observed before—but increases with higher attraction towards beacons. Indeed, a high value of σ_b makes all UAVs move towards the weed patches first, leaving other areas of the field unattended and resulting in a overall higher coverage time, as the field gets fully covered only when all cells receive at least one visit. Note that coverage times are slightly higher in this case than what is shown in Fig. 3, because the mapping task requires multiple passages over the same cell, hence slowing down coverage. On the other hand, the mapping time t_m decreases with high attraction towards beacons, as more UAVs are dedicated to mapping only those areas that require attention. Conversely, mapping is less efficient when repulsion among UAVs is too strong, to the point that no substantial difference with coverage is visible when $\sigma_a = 64$. The smallest values of t_m occur for medium values of attraction and large values of repulsion ($\sigma_a = 8$ and $\sigma_b = 32$ in Fig. 4). These values slightly vary with the group size N, as shown in the appendix, but generally indicate that there can be positive interactions between repulsion among UAVs and attraction towards beacons.

---

[2] Data for different group sizes are available in the appendix at the end of the manuscript.

When some error in perception is introduced with artificial disturbances on the images (bottom panels in Fig. 4), the coverage and mapping time in general increases due to the need to frequently revisit those cells where some perception error occurred. Such negative effects are negligible for what concerns the coverage time, because the entire field must be covered in any case. The mapping time gets instead much worse, due to the need to visit multiple times just those cells where weed is present. For $\sigma_a = 64$, mapping terminates even after coverage, indicating that repulsion among agents is too strong to make UAVs focus on the weed infested areas en masse.

The accuracy of mapping in presence of perception error increases thanks to the multiple visits performed to weed-infested areas, which allow to increase the probability of detecting the correct number of balls, as UAVs visiting a cell at different times get a different perception error. The detection error decreases below 5% (mean: 0.046, standard deviation: 0.009), indicating that collaboration among UAVs is effective to increase mapping accuracy.

3.3   Hardware-in-the-Loop (HIL) Simulations

Simulations have been performed also to profile the developed algorithm when run on the RPi of the UAV. One drone has been connected through the serial port to a desktop, and interacted with the simulator through UDP messages. The UAV process onboard images from the dataset corresponding to the simulated cells, and decides the next cell to visit. The new waypoint is not sent to the autopilot but rather it is communicated to the simulator that implements the UAV motion. Similarly, ORCA is executed onboard and new waypoints are generated and sent to the simulator. We have performed several profiling tests to understand how much time is required for each operation. Overall, the module that takes longer time is `Perception`, which takes approximately 0.244 s in average, while `Core` takes about 0.107 s. Considering that `Perception` is executed only once per cell while the UAV is hovering, these values are compatible with field deployment, confirming that the proposed strategy can be reliably tested with real UAVs.

4   Conclusions

With this study, we have moved a fundamental step in the direction of field deployment for UAV swarms. We have described an efficient implementation on a real platform of a scalable coverage and mapping strategy based on reinforced random walks. The navigation strategy simplifies and improves over previous work [2], including collision avoidance among UAVs and also enhancing the random walk. The latter has been modified to take into account both the direction and the intensity of the bias vector resulting from attraction to beacons and repulsion from other agents. In this way, all available information are exploited for the benefit of both field coverage and weed mapping. In computing the results, we put particular care in the realism of the simulation, including artificial vision. Additionally, we implemented the proposed strategy on the real UAV hardware and tested it into hardware-in-the-loop simulations, verifying the suitability of the implementation prior to deployment with flying UAVs. Proof-of-concept videos of the real flying system is available online [16]. Our current effort is in collecting evidence of the suitability of the proposed strategy with more field tests, to be performed both indoor, using the mockup experimental scenario described in this paper, and outdoor, with tests performed on agricultural fields and real onboard classification of corps and weeds. The latter proves particularly challenging, due to the complexity of the vision routines that are prone to non-negligible errors. However, this paper suggests that swarms of UAVs can improve the detection accuracy beyond the individual limitation, and field tests will be dedicated to support and strengthen this concept. We are already moving the next steps toward future work, specifically by introducing a Bayesian estimator to reliably determine if a cell can be considered mapped or not. This is then used by the RRW to improve the exploration strategy based on the information that a specific region is expected to provide.

Acknowledgments. This work has been supported by SAGA (Swarm Robotics for Agricultural Applications), an experiment founded by the European project ECHORD++ (GA: 601116). Dario Albani and Daniele Nardi acknowledge partial support from the European project FLOURISH (GA: 644227).

A   Appendix - Additional Experiments

In this section we present additional data coming from some more experiments performed within this study. In particular, we present results for coverage and mapping time obtained by varying the number of robots involved in the simulation. As expected, both the mapping and the coverage problem benefit from the increased density of agents. We also observe that such increase in performance does not scale linearly due to non-beneficial interactions between the agents (i.e. issues related to overcrowding). Last, we observe a "right shift" in the minima of the mapping time t_m when the number of agents in the swarm increases. This is expected, as a larger number of agents increases the repulsion force acting on the single UAV, thus requiring higher attraction forces from the beacons to be effective (Figs. 5, 6 and 7).

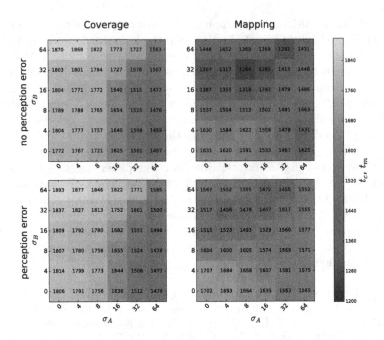

Fig. 5. Comparison of the coverage and mapping time for N = 50 robots. Each cell in the heatmap represents the average of 150 runs.

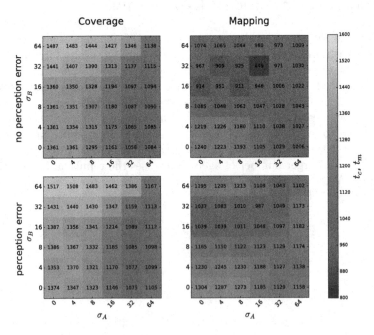

Fig. 6. Comparison of the coverage and mapping time for N = 75 robots. Each cell in the heatmap represents the average of 150 runs.

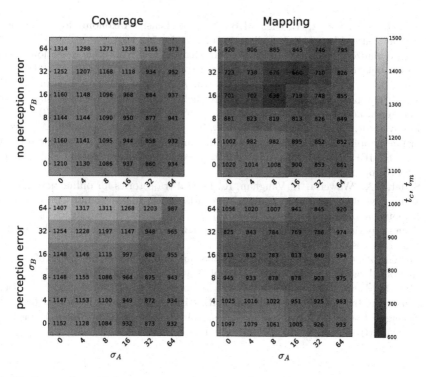

Fig. 7. Comparison of the coverage and mapping time for N = 100 robots. Each cell in the heatmap represents the average of 150 runs.

References

1. Albani, D., Manoni, T., Nardi, D., Trianni, V.: Dynamic UAV swarm deployment for non-uniform coverage. In: AAMAS 2018: Proceedings of the 2018 International Conference on Autonomous Agents and Multiagent Systems, pp. 1–9 (2018)
2. Albani, D., Nardi, D., Trianni, V.: Field coverage and weed mapping by UAV swarms. In: 2017 IEEE/RSJ International Conference on Intelligent Robots and Systems (IROS), pp. 4319–4325. IEEE (2017)
3. Avular. https://www.avular.com. Accessed 22 Apr 2018
4. Bareiss, D., van den Berg, J.: Generalized reciprocal collision avoidance. Int. J. Robot. Res. **34**(12), 1501–1514 (2015)
5. Brambilla, M., Ferrante, E., Birattari, M., Dorigo, M.: Swarm robotics: a review from the swarm engineering perspective. Swarm Intell. **7**(1), 1–41 (2013)
6. Field robot event. http://www.fieldrobot.com/event/. Accessed 22 Apr 2018
7. Gerkey, B.P., Matarić, M.J.: A formal analysis and taxonomy of task allocation in multi-robot systems. Int. J. Robot. Res. **23**(9), 939–954 (2004)
8. Hoffmann, H., Jensen, R., Thomsen, A., Nieto, H., Rasmussen, J., Friborg, T.: Crop water stress maps for an entire growing season from visible and thermal UAV imagery. Biogeosciences **13**(24), 6545–6563 (2016)
9. King, A.: Technology: the future of agriculture. Nature **544**(7651), 21–23 (2017)
10. Koeveringe, M., van Evert, F., Li, Y., Kootstra, G.: Detection of broad-leaved weed plants in grasslands, (in preparation)

11. Korsah, G.A., Stentz, A., Dias, M.B.: A comprehensive taxonomy for multi-robot task allocation. Int. J. Robot. Res. **32**(12), 1495–1512 (2013)
12. Nieuwenhuizen, A.T., Hofstee, J.W., van Henten, E.J.: Adaptive detection of volunteer potato plants in sugar beet fields. Precis. Agric. **11**(5), 433–447 (2009)
13. Peña, J.M., Torres-Sánchez, J., de Castro, A.I., Kelly, M., López-Granados, F.: Weed mapping in early-season maize fields using object-based analysis of unmanned aerial vehicle (UAV) images. PLoS ONE **8**(10), e77151 (2013)
14. Popović, M., Vidal-Calleja, T., Hitz, G., Sa, I., Siegwart, R., Nieto, J.: Multiresolution mapping and informative path planning for UAV-based terrain monitoring. In: 2017 IEEE/RSJ International Conference on Intelligent Robots and Systems (IROS), pp. 1382–1388. IEEE (2017)
15. Sadat, S.A., Wawerla, J., Vaughan, R.: Fractal trajectories for online non-uniform aerial coverage. In: Proceedings of the 2015 IEEE International Conference on Robotics and Automation (ICRA 2011), pp. 2971–2976. IEEE (2015)
16. Saga experiment media center. http://laral.istc.cnr.it/saga/index.php/media-center. Accessed 22 Apr 2018
17. Stevens, A., Othmer, H.: Aggregation, blowup, and collapse: the ABC's of taxis in reinforced random walks. SIAM J. Appl. Math. **57**(4), 1044–1081 (1997)
18. Van Den Berg, J., Guy, S.J., Lin, M., Manocha, D.: Reciprocal n-body collision avoidance. In: Pradalier, C., Siegwart, R., Hirzinger, G. (eds.) Robotics Research. Springer Tracts in Advanced Robotics, vol. 70, pp. 3–19. Springer, Heidelberg (2011). https://doi.org/10.1007/978-3-642-19457-3_1

Self-Assembly from a Single-Molecule Perspective

Kevin R. Pilkiewicz[1]([✉]), Pratip Rana[2], Michael L. Mayo[1],
and Preetam Ghosh[2]

[1] U.S. Army Engineer Research and Development Center, Vicksburg, MS 39180, USA
Kevin.R.Pilkiewicz@usace.army.mil
[2] Department of Computer Science, Virginia Commonwealth University, Richmond,
VA 23284, USA

Abstract. As manipulating the self-assembly of supramolecular and nanoscale constructs at the single-molecule level increasingly becomes the norm, new theoretical scaffolds must be erected to replace the thermodynamic and kinetics based models used to describe traditional bulk phase active syntheses. Like the statistical mechanics underpinning these latter theories, the framework we propose uses state probabilities as its fundamental objects; but, contrary to the Gibbsian paradigm, our theory directly models the transition probabilities between the initial and final states of a trajectory, foregoing the need to assume ergodicity. We leverage these probabilities in the context of molecular self-assembly to compute the overall likelihood that a specified experimental condition leads to a desired structural outcome. We demonstrate the application of this framework to a simple toy model in which three identical molecules can assemble in one of two ways and conclude with a discussion of how the high computational cost of such a fine-grained model can be overcome through approximation when extending it to larger, more complex systems.

Keywords: Molecular communication · Soft matter · Statistical mechanics

1 Introduction

The concept of structure-driven communication is first grasped at a very young age, when one first puzzles out that the square peg must be inserted into the square hole. But at a much earlier stage of development, this "lock and key" motif is already ingrained into our bodies on the microscale, where evolution has engineered countless proteins whose native states form a pocket that is sized and shaped to bind only a single, specific molecular partner. In the macroworld, we make locks that admit only a single key shape so that we may exclude others from our privacy. Exclusivity is the objective of structural communication in the

© ICST Institute for Computer Sciences, Social Informatics and Telecommunications Engineering 2019
Published by Springer Nature Switzerland AG 2019. All Rights Reserved
A. Compagnoni et al. (Eds.): BICT 2019, LNICST 289, pp. 147–155, 2019.
https://doi.org/10.1007/978-3-030-24202-2_11

microworld as well, though it is not volitional interlopers who are the concern but rather stochastic ones. In the noisy environment of our cells, structural exclusion is the only way to ensure that a protein does not bind the first molecule to diffuse into it.

While structural specificity may combat the stochasticity inherent to molecular binding, the stochasticity inherent to the self-assembly of these structures themselves is a much higher hurdle to surpass. Even nature, with four billion years of evolutionary experience, has not perfected this art. The misfolding of proteins like the neuronal amyloid-beta ($A\beta$) protein or the pancreatic amylin, for example, can seed the formation of plaques that have been implicated as a potential cause for Alzheimer's disease and type II diabetes, respectively [1]. Human efforts to synthesize nanoscale structures that can interact with or leverage biology have thus understandably struggled with precise structural control. Gold nanoparticle [2] and liposome [3] syntheses have difficulty achieving acceptable levels of monodispersity, biofilm [4] and other monolayer surface depositions [5] are prone to disorder and defects, and supramolecular assemblies [6] are often plagued by competing interactions that lead to disparate products.

Newer techniques such as optical and magnetic tweezers [7], molecular beams [8], and micro- and nanofluidics [9] have shown potential for greatly improving our control over molecular self-assembly processes by reducing the scale of the experiments from the macroscopic bulk phase to systems involving only a few relevant molecules. The current theories used to model self-assembling systems, however, still largely rely on bulk statistical thermodynamics and kinetics [10–12], which are insufficient for this new experimental scale. In this paper we attempt to address this gap by modeling the distribution of self-assembled states by considering the stochastic dynamics of a single self-assembly sequence and its branching structural end states. The result is a probabilistic model that requires neither an ergodic hypothesis nor a thermodynamic limit.

2   The Model

The framework of our model is to assign to each molecule two sets of stochastic variables. The first set, which we denote as the state ψ, characterizes the molecule's entry into the system. This set of parameters might include the time at which it is injected or emitted, its initial position and velocity, and its starting orientation. These are parameters over which the experimenter exercises some degree of control. The second set, which we denote by the state ϕ, describes the molecule's interaction with the self-assembling core. Whether or not the molecule adds to the growing structure and in what manner it adds will depend upon the same sorts of parameters, but evaluated at the time of first interaction, which itself may be one of the random variables in ϕ. We denote the set of all possible end states of the self-assembly process under consideration as S, and the probability that self-assembly terminates at some structure $s \in S$ will depend upon the states ϕ of each interacting molecule—even those that interacted without binding to the structure. We call this set of states $\{\phi\}$ and the probability linking this set to a specified outcome as $p_R(s \in S|\{\phi\})$, where the R index stands for "result."

The set of states $\{\phi\}$ evolve dynamically from the set of initial states $\{\psi\}$ as a result of some stochastic transport process. This process might be simple diffusion through a volume or across a surface, or it might be some facilitated process. The probability of observing a specific set of interacting states given a set of initial states is defined as the conditional transport probability $p_T(\{\phi\}|\{\psi\})$. We further define the source probability $p_S(\{\psi\}|\sigma)$ as the likelihood that a set of initial states $\{\psi\}$ are observed given specified values of a set of externally tunable parameters σ. This set can consist of variables like temperature and emission frequency that are directly manipulated by the experimenter. The ultimate probability that we wish to compute is $p_F(s|\sigma)$, the overall final probability that a set of input parameters will result in molecules assembling into structure s. This probability can be related to the latter three by the following integral:

$$p_F(s|\sigma) = \int d\{\phi\}d\{\psi\}\, p_R(s|\{\phi\})p_T(\{\phi\}|\{\psi\})p_S(\{\psi\}|\{\sigma\}). \tag{1}$$

The complexity involved in actually evaluating Eq. (1) will naturally depend upon the details of the system under consideration. In this paper, we restrict our attention to a simple toy model that demonstrates how this theoretical framework might be applied and what sort of predictions it can be used to make. In this model, we assume that three identical molecules are released at randomly selected times into a one-dimensional drift-diffusion channel characterized by a drift speed v, a diffusion constant D, and a channel length ℓ. The first molecule to traverse the channel binds to a receptor site that catalyzes a self-assembly process with the second molecule to arrive, resulting in a dimer state. We assume that this assembly process takes a finite amount of time, which we denote as the assembly time T_α. If the third molecule arrives while dimer assembly is still occurring, it will be repelled and the final state of the system will be the dimer. If, on the other hand, it arrives once the dimer is complete, then a trimer state will result. Figure 1 depicts a cartoon representation of our toy model and summarizes its possible outcomes.

If the three molecules are labeled 0, 1, and 2 based on the order in which they are released into the channel, we may define the initial state of the ith particle ψ_i as its injection time τ_i, and its interaction state ϕ_i can be defined analogously as its arrival time t_i at the channel terminus. Because the first such arrival time may be thought of as the start of the experiment, the absolute release and arrival times are irrelevant, and we can replace these six time variables with four time intervals. We define the intervals $\Delta\tau_1$ and $\Delta\tau_2$ as the differences between the injection times of particle 1 and particle 0 and particle 2 and particle 0, respectively. Analogously, the intervals Δt_1 and Δt_2 are the equivalent differences in the arrival times of the particles at the self-assembly site. Because the order in which the particles arrive is not fixed, due to the stochasticity of the transport down the channel, these latter time intervals can potentially be negative. The space of self-assembled structures S in this case contains only the dimer and trimer configurations, which we shall denote as s_2 and s_3, respectively. The probability of observing a dimer at the end of the experiment depends upon whether or not the third molecule to interact does so within a time T_α of the second molecule's arrival.

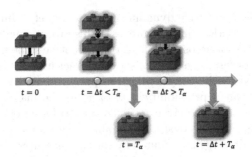

Fig. 1. Toy model timeline. If $t = 0$ is the time at which the first two monomers initiate their self-assembly into a dimer, then $t = \Delta t$ is the time at which the third monomer arrives. Depending upon whether this time is smaller or larger than the self-assembly time scale T_α, the final state of the model will either be the original dimer or a trimer.

We can thus write the dimer result probability $p_R(s_2|\Delta t_1, \Delta t_2)$ as the following conditional:

$$p_R(s_2|\Delta t_1, \Delta t_2) = \begin{cases} 1 - \Theta(|\Delta t_2 - \Delta t_1| - T_\alpha) & \text{for } \Delta t_1 \geq 0, \Delta t_2 \geq 0 \\ 1 - \Theta(|\Delta t_2| - T_\alpha) & \text{for } \Delta t_2 \geq \Delta t_1, \Delta t_1 < 0 \quad (2) \\ 1 - \Theta(|\Delta t_1| - T_\alpha) & \text{for } \Delta t_1 \geq \Delta t_2, \Delta t_2 < 0. \end{cases}$$

In the above, $\Theta(t)$ is the Heaviside step function. We adopt the convention that it takes value unity when its argument exceeds zero and has value zero otherwise. Since the only other possibility is that a trimer is formed, $p_R(s_3|\Delta t_1, \Delta t_2) = 1 - p_R(s_2|\Delta t_1, \Delta t_2)$.

The first passage time across a drift-diffusion channel is distributed according to the standard inverse Gaussian distribution $IG(\mu, \lambda; t)$ [13], analytically continued to be zero for negative values of its time argument:

$$IG(\mu, \lambda; t) = \begin{cases} \sqrt{\frac{\lambda}{2\pi t^3}} \exp\left[\frac{-\lambda(t-\mu)^2}{2\mu^2 t}\right] & t > 0 \\ 0 & t \leq 0 \end{cases} \quad (3)$$

The parameter $\mu \equiv \ell/v$ is the time it takes to cross the channel in the absence of diffusion, and $\lambda \equiv \ell^2/2D$ is the average time it would take in the absence of drift. This suggests the following form for the transport distribution $p_T(\Delta t_1, \Delta t_2|\Delta \tau_1, \Delta \tau_2)$:

$$p_T(\Delta t_1, \Delta t_2|\Delta \tau_1, \Delta \tau_2)$$
$$= \int_0^\infty dt\, IG(\mu, \lambda; t) IG(\mu, \lambda; t + \Delta t_1 - \Delta \tau_1)$$
$$\times IG(\mu, \lambda; t + \Delta t_2 - \Delta \tau_2). \quad (4)$$

Finally, we will assume that each molecule has an equal chance of being released into the channel at any moment in time after the previous molecule's emission, which results in the release time intervals being exponentially distributed (as in a radioactive decay process). Assuming an average injection rate $1/\tau$, we get

$$p_S(\Delta\tau_1, \Delta\tau_2|\tau) = \frac{1}{\tau^2}e^{-(\Delta\tau_2 - \Delta\tau_1)/\tau}e^{-\Delta\tau_1/\tau}$$

$$= \frac{1}{\tau^2}e^{-\Delta\tau_2/\tau}. \tag{5}$$

Note that the dependence of this distribution on $\Delta\tau_1$ cancels out of the exponent and that τ is the presumptive tuning parameter of the experiment.

Equations (2), (4), and (5) can be substituted into Eq. (1) to calculate the total dimer probability $p_F(s_2|\tau)$. The simple conditional form of the result probability $p_R(s_2|\Delta t_1, \Delta t_2)$ will lead to a modification in the limits of integration over the arrival time intervals. This leads to a more complicated looking expression for $p_F(s_2|\tau)$ that is nonetheless more straightforward to evaluate numerically:

$$p_F(s_2|\tau) = \frac{1}{\tau^2}\int_0^\infty d\Delta\tau_2 \int_0^\infty d\Delta\tau_1 \, e^{-\Delta\tau_2/\tau}$$

$$\times \left[2\int_0^\infty d\Delta t_2 \int_{\max(\Delta t_2 - T_\alpha, 0)}^{\Delta t_2} d\Delta t_1 \, p_T(\Delta t_1, \Delta t_2|\Delta\tau_1, \Delta\tau_2) \right.$$

$$+ 2\int_0^{T_\alpha} d\Delta t_2 \int_0^\infty d\Delta t_1 \, p_T(-\Delta t_1, \Delta t_2|\Delta\tau_1, \Delta\tau_2)$$

$$\left. + 2\int_0^{T_\alpha} d\Delta t_2 \int_{\Delta t_2}^\infty d\Delta t_1 \, p_T(-\Delta t_1, -\Delta t_2|\Delta\tau_1, \Delta\tau_2) \right]. \tag{6}$$

The three integrals over the arrival time intervals in the above expression correspond, respectively, to the cases in which particle 0 arrives first, second, and third. The factors of 2 account for the symmetry, in each case, of swapping the index labels 1 and 2.

3   Results and Discussion

Even for such a simple toy system, the numerical integration required to calculate $p_F(s_2|\tau)$ is computationally intensive, with the principal time sink being the repeated evaluations of Eq. (4) for all the different values of the release and arrival time intervals needed to evaluate Eq. (6). We resolved this difficulty by parallelizing the computation, evaluating each instance of $p_T(\Delta t_1, \Delta t_2|\Delta\tau_1, \Delta\tau_2)$ on a separate thread of an Nvidia GeForce GTX TITAN GPU with 3,072 cores, 12 GB of RAM, and 1,000 MHz clock speed. This reduced the total computational time by a factor of roughly 1,000. For our integration mesh, we chose a lattice spacing (bin width) of 0.02 time units, and a mesh domain defined in terms of model time units by the inequalities $0 \leq \Delta\tau_1 \leq \Delta\tau_2$, $0 \leq \Delta\tau_2 \leq r$, $-10 \leq \Delta t_1 \leq r + 10$, and $-10 \leq \Delta t_2 \leq r + 10$, where the integration range r was set equal to 20 time units. For each point on this mesh, the formally infinite upper limit of each of the parallelized time integrals was approximated as 50 time units. These restricted integration ranges were sufficient to approximately normalize all of the probability distributions of the model to within an acceptable tolerance.

After computing the transport probability at each point of the chosen integration mesh, it became tractable to evaluate the integrals over the release and arrival time intervals serially, using an Intel Core i7-2600 CPU with 3.40 GHz clock speed and 8 GB of memory. We demonstrate how this computational time varies with integration range r in Fig. 2 for three different bin widths. As the logarithmic scale makes clear, the serial computation time grows roughly exponentially with the integration range. It also grows approximately as an inverse power law of the bin width, with a negative exponent of about 4.

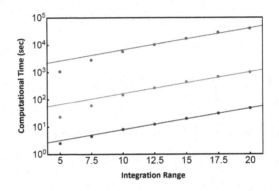

Fig. 2. Computational time (in seconds) plotted versus the integration range (in model time units) for three different bin widths (also in model time units): from top to bottom, 0.02 (red), 0.05 (green), and 0.1 (blue). The ordinate axis is on a log scale to better illustrate the exponential growth of the computational time for sufficiently large integration ranges. (Color figure online)

We plot our numerically evaluated probability $p_F(s_2|\tau)$ in Fig. 3(A) as a function of the self-assembly time scale T_α for values of the mean release interval $\tau = 0.5, 1.5, 2.5,$ and 3.5, in descending order. For all curves, the time scales μ and λ were both fixed at unity. As expected, when self-assembly is instantaneous ($T_\alpha = 0$), there is no interval of time during which the third molecule can be repelled, so trimer formation is inevitable ($p_F(s_2|\tau) = 0$). At the other extreme, as $T_\alpha \to \infty$, the dimer becomes the only possible product ($p_F(s_2|\tau) \to 1$). As molecule emissions into the channel become more infrequent (larger τ), the window to avoid trimer trapping becomes smaller, depressing the dimer probability. These curves are all fit very well by a function of the form $1 - \exp[-c_1(T_\alpha/\tau)^{c_2}]$, where c_1 and c_2 are fitting parameters that may depend in a complicated manner upon some dimensionless combination of the time scales τ, μ, and λ. These best fit functions are plotted as solid curves over the numerical data in Fig. 3(A).

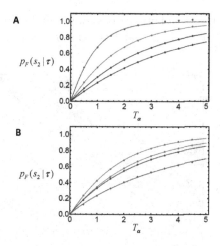

Fig. 3. Plots of the final dimer formation probability $p_F(s_2|\tau)$ versus the self-assembly time scale T_α for (A) fixed μ and λ equal to unity, with τ varying, from top to bottom, as 0.5, 1.5, 2.5, and 3.5; and for (B) fixed $\tau = 1.5$ with (μ, λ) varying, from top to bottom, as $(1,1)$, $(2,2)$, $(2,1)$, and $(4,2)$. The data points are the computationally evaluated probabilities, while the solid curves are the analytic fits of the function $1 - \exp[-c_1(T_\alpha/\tau)^{c_2}]$.

Figure 3(B) also plots the final dimer probability versus T_α, but this time τ is held fixed at $\tau = 1.5$ and μ and λ are varied instead. The top curve is the same as the second to top curve in panel (A) ($\mu = \lambda = 1$). The remaining curves are, in descending order, for $(\mu, \lambda) = (2, 2)$, $(2, 1)$, and $(4, 2)$. These curves illustrate several general trends. First, the dimer probability decreases monotonically with increasing μ, reflecting the fact that a less facilitated channel will tend to space out the arrival times of the molecules, making trimer formation more likely. Increasing λ tends to have the opposite effect, since reducing the diffusivity of the channel narrows the distribution of arrival times (Eq. (3)), resulting in a less noisy channel. The variance of the inverse Gaussian distribution is μ^3/λ, explaining why $p_F(s_2|\tau)$ has a stronger dependence on μ than on λ. These curves are well modeled by the same class of fitting function used in Fig. 3(A).

Perhaps the most informative way of quantifying how self-assembly depends upon our model parameters is with a "phase diagram," where a relevant parameter subspace is divided into regions based upon the most probable structure in each. For our toy system, this phase diagram is fairly simple and is plotted in Fig. 4 as a function of the control parameter τ and the self-assembly time T_α. The transport parameters μ and λ are both fixed at unity. The phase boundary, which turns out to be approximately linear (R^2-value of ≈ 0.997), was determined by finding, for each value of τ, the critical value of T_α for which $p_F(s_2|\tau) = 1/2$. For large T_α and small τ, the shorter average interval between particle emissions and the longer assembly time will make it more likely for the third particle to arrive while the first two are still docking, thereby frustrating trimer formation.

In the opposite limit, the time between emissions will be long and assembly will occur swiftly–both circumstances that favor the trimer product.

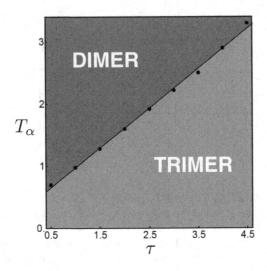

Fig. 4. The phase diagram for the toy model when $\mu = \lambda = 1$. The phase boundary separating the dimer and trimer favoring regions is approximately linear.

4   Conclusions

While the framework we have devised for quantifying self-assembly in terms of individual molecular interactions is quite general, we have seen even in the simple case of our three-molecule toy model that its computational cost is problematic, especially were one to extend it to the self-assembly of long biopolymers like proteins or microtubules. The set of interaction variables $\{\phi\}$ will necessarily grow linearly with the number of interactions considered, but the real problem is that an integral like that in Eq. (4) will have to be evaluated for every permissible set of values these variables can take. The number of integrations will thus grow exponentially with the number of interactions, rendering even parallelization schemes unfeasible for supramolecular assemblies consisting of more than a handful of subunits.

The most straightforward way to address this problem is to make physically sensible approximations that constrain the hypervolumetric domain of the variables $\{\phi\}$, thereby reducing the number of integrals that must be computed in parallel. In our toy model, for example, we must consider a range of Δt_1 and Δt_2 values broad enough to allow for $3! = 6$ different interaction orders. If we work in the large τ limit, however, we can assume that the probability of nonconsecutively released particles interacting in reversed order is negligible. This reduces the number of permissible interaction orderings to three (removing the factor of two from the second term on the right of Eq. (6) and deleting the third term entirely) and eliminates the need to consider negative values of Δt_2. This is only a modest gain, but if we were to extend our toy model to include tetrameric structures, this approximation scheme would reduce the number of allowable orderings from $4! = 24$ to a paltry five.

Our intention moving forward is to begin exploring the impact these sorts of approximation schemes have on both computational time and numerical accuracy in the hopes of ultimately applying the methodology outlined in this paper to systems of actual biological consequence.

Acknowledgments. The work described in this document was funded under the U.S. Army Basic Research Program under PE 61102, Project T25, Task 02 "Network Science Initiative" and was managed and executed at the U.S. Army ERDC. Opinions, interpretations, conclusions, and recommendations are those of the author(s) and are not necessarily endorsed by the U.S. Army.

References

1. Knowles, T.P.J., Vendruscolo, M., Dobson, C.M.: The amyloid state and its association with protein misfolding diseases. Nat. Rev. Mol. Cell Biol. **15**, 384–396 (2014)
2. Chatterjee, S., Bandyopadhyay, A., Sarkar, K.: Effect of iron oxide and gold nanoparticles on bacterial growth leading towards biological application. J. Nanobiotechnol. **9**(34), 1–7 (2011)
3. Jing, Y., Trefna, H., Persson, M., Kasemo, B., Svedhem, S.: Formation of supported lipid bilayers on silica: relation to lipid phase transition temperature and liposome size. Soft Matter **10**, 187–195 (2014)
4. Jin, X., Riedel-Kruse, I.H.: Biofilm Lithography enables high-resolution cell patterning via optogenetic adhesin expression. PNAS **115**(14), 3698–3703 (2018)
5. Jang, J., Hong, S., Schatz, G.C., Ratner, M.A.: Self-assembly of ink molecules in dip-pen nanolithography: a diffusion model. J. Chem. Phys. **115**(6), 2721–2729 (2001)
6. Chakrabarty, R., Mukherjee, P.S., Stang, P.J.: Supramolecular coordination: self-assembly of finite two- and three-dimensional ensembles. Chem. Rev. **111**, 6810–6918 (2011)
7. Biancaniello, P.L., Kim, A.J., Crocker, J.C.: Colloidal interactions and self-assembly using DNA hybridization. Phys. Rev. Lett. **94**, 058302 (2005)
8. Xin, S.H., et al.: Formation of self-assembling CdSe quantum dots on ZnSe by molecular beam epitaxy. Appl. Phys. Lett. **69**, 3884 (1996)
9. Jahn, A., Vreeland, W.N., Gaitan, M., Locascio, L.E.: Controlled vesicle self-assembly in microfluidic channels with hydrodynamic focusing. J. Am. Chem. Soc. **126**, 2674–2675 (2004)
10. Israelachvili, J.N., Mitchell, D.J., Ninham, B.W.: Theory of self-assembly of hydrocarbon amphiphiles into micelles and bilayers. J. Chem. Soc. Faraday Trans. 2 **72**, 1525–1568 (1976)
11. Nagarajan, R., Ruckenstein, E.: Theory of surfactant self-assembly: a predictive molecular thermodynamic approach. Langmuir **7**, 2934–2969 (1991)
12. Sweeney, B., Zhang, T., Schwartz, R.: Exploring the parameter space of complex self-assembly through virus capsid models. Biophys. J. **94**, 772–783 (2008)
13. Folks, J.L., Chhikara, R.S.: The inverse Gaussian distribution and its statistical application-a review. J. R. Stat. Society. Ser. B (Methodol.) **40**(3), 263–289 (1978)

Cyber Regulatory Networks: Towards a Bio-inspired Auto-resilient Framework for Cyber-Defense

Usman Rauf[1]([✉]), Mujahid Mohsin[2], and Wojciech Mazurczyk[3]

[1] Department of Software and Information Systems,
University of North Carolina at Charlotte, Charlotte, NC, USA
urauf@uncc.edu, usman.cyberdna@gmail.com
[2] National University of Sciences and Technology, Islamabad, Pakistan
mmohsin@cae.nust.edu.pk
[3] Institute of Telecommunications, Warsaw University of Technology,
Warsaw, Poland
wmazurczyk@tele.pw.edu.pl

Abstract. After decades of deploying cyber-security systems, it has become a well-known fact that the existing cyber-security architecture has numerous inherent limitations that make the maintenance of the current network security devices unscalable and provide the adversary with asymmetric advantages. These limitations include: (1) difficulty in obtaining the global network picture due to lack of mutual interactions among heterogeneous network devices, (2) poor device self-awareness in current architectures, (3) error-prone and time consuming manual configuration which is not effective in real-time attack mitigation, (4) inability to diagnose misconfiguration and conflict resolution due to multi-party management of security infrastructure. In this paper, as an initial step to deal with these issues, we present a novel bio-inspired auto-resilient *security* architecture. The main contribution of this paper includes: (1) investigation of laws governing the dynamics of correct feedback control in Biological Regulatory Networks (BRNs), (2) studying their applicability for synthesizing correct models for bio-inspired communication networks, i.e. Firewall Regulatory Networks (FRNs), (3) verification of the formal models of real network scenarios, to prove the correctness of the proposed approach through model checking techniques.

1 Introduction

With the ever increasing number of data breaches and security incidents it is evident that the traditional manual models of cyber-security are unable to defend complex and large cyber-networks. The new models of defence need to focus on auto-resiliency, integration and fast response-time. To meet these objectives, even after decades of development of cyber security systems, still there exist

© ICST Institute for Computer Sciences, Social Informatics and Telecommunications Engineering 2019
Published by Springer Nature Switzerland AG 2019. All Rights Reserved
A. Compagnoni et al. (Eds.): BICT 2019, LNICST 289, pp. 156–174, 2019.
https://doi.org/10.1007/978-3-030-24202-2_12

inherent limitations in the current cyber-security architecture that allow adversaries to not only plan and launch attacks effectively but also learn and evade detection quite easily. These limitations include: (1) difficulty to obtain the global knowledge (about security policies) of all security devices in the network, e.g., different firewalls are often managed by independent administrators, who have no incentive for conducting coordinated exercises to mitigate the global risk, (2) manual reconfiguration, which is time consuming, error-prone and ineffective in real-time attack mitigation, (3) lack of mutual interactions among network devices as most security devices such as firewalls and IDSs are configured and managed individually for a subset of assets being directly protected, without any realization of the global impact caused by reconfiguration of a single device on the entire enterprize, (4) multi-party management makes the diagnosis (for misconfiguration) and conflict resolution difficult and often leads to un-optimized policy. Current cyber-security architecture does not have any notion of coordination via interaction among security devices. The limited sensing mechanisms that exist work mostly offline, require human assistance and do not converge towards global optimal solution. Our proposed work is inspired by interaction at the cellular level between the entities of a Biological Regulatory Network (BRN), which is a very fundamental and crucial phenomenon in the dynamics of biological systems [25].

Existing cyber-security architectures also have no self-awareness of the risk as the security policies usually take into account only usability, reachability and demand requirements; whereas, the risk realization is either partially or completely ignored. Although the state of the art security risk assessment frameworks provide a general overview of assessing and mitigating threats in different phases of a kill-chain [1,28], these frameworks do not provide any means for aligning local goals with global objectives. Interaction and self-awareness are the fundamental ingredients for self-organization and adaptivity. Note that sensing and self-awareness are two completely different processes, as sensing is related to observing one's environment or neighbors through interaction, and self-awareness is the realization of one's internal state. The sensing process without the aim of optimizing a global objective (via feedback notion and automation) is nearly useless, as by the time humans take action, the damage could be already done, or becomes uncontrollable. The hierarchy in current cyber architectures lacks these functionalities. Therefore, there is an immense need to integrate feedback mechanism in current architectures to allow continuous and dynamic risk mitigation and real time response (in case of any perturbation).

On the other hand, biological systems have built-in feedback mechanisms, through which they adapt and survive unknown threats in the surrounding environment. We intend to redesign cyber-security architecture, based upon such regulatory, and feedback control mechanisms, in which even if one router, device or machine is compromised, the neighboring devices should have tendency to alter their behavior by allowing or restricting it from performing malicious activities. The resultant architecture should have the tendency to survive under abnormal conditions and to reduce (global) risk factor by maintaining progressive cycle to avoid a deadlock/malicious state where risk is higher/above a specific threshold.

1.1 Challenges

The incentive of every security device in a cyber infrastructure is to reduce the risk and to increase the usability and demand of the assets, for which it is responsible. In large scale networks many security devices are intertwined, and security policies of any device are not designed to reinforce neighboring security devices, rather they are more focused towards the interests (usability and demand) of the important assets that they are protecting. Consequently, a single erroneous action performed by an operator (to fulfill demand) at a local level in any security device, might have catastrophic (infrastructure level) impact (by increasing risk on the other devices), which is hard (or impossible) to comprehend via manual configuration. To the best of our knowledge, no existing technique implements correct feedback control mechanism, for reconciliation among security devices, to automate the global risk mitigation in an infrastructure. Designing optimal policies to mitigate risk at global level is a Distributed Constraint Optimization Problem (DCOP) [20], for which time complexity is exponential in the worst case scenario, and managing such sensitive system manually is nearly impossible.

1.2 Contributions

As a first step towards creating an auto-resilient cyber architecture, in this paper we present a three fold contribution: (1) first, we investigate the laws governing the dynamics of correct feedback control in BRNs, (2) then we apply these laws to synthesize correct model for bio-inspired networks, (3) finally we verify the synthesized models for real communication networks, through model checking techniques, to prove the correctness of the proposed approach.

2 Related Work

Over the past few years, bio-inspired computing has evolved as an active area of research. Different aspects of biological phenomenon give rise to bio-inspired mechanisms with applications to cyber-security, which include: (1) Swarm Intelligence (SI), (2) Artificial Immune system (AIS), (3) Genetic Mutation, and (4) Gene and Cell Regulation.

SI can be defined as an emergent collective behavior of non intelligent interacting entities that attempt to achieve self-organization and global objectives without any centralized control. Inspired by the behavior of social insects, the main focus of the domain is to design resilient and robust systems, which can efficiently and intelligently operate under the threat and catastrophic conditions without any centralized control [3]. Mostly, SI based approaches have been used for efficient routing, for identifying the source of an attack in the network, i.e. Intrusion Detection System (IDS), and to prevent the attack by localizing its origin, i.e. Intrusion Prevention System (IPS). Few recent approaches towards this direction are [12,14,19,21,27].

The research in the domain of AIS began in the mid-1980s with Farmer, Packard, and Perelson's study [11]. The biggest revolution in this domain was the utilization of the concept of human immune system for computer security which proposed one to one mapping or analogy between the immune system and IDSes. With the development of the HIS principle, Negative Selection Algorithm (NSA) [13], Clonal Selection Algorithm (CSA) [5], Immune Network Algorithm (INA) [4], and Danger Theory Algorithm (DTA) [2] become the most representative algorithms in this domain. The most recent approaches which utilize these concepts to design or improve IDSes includes [17, 18, 30].

Recently, researchers from the domain of cyber security have mapped the concept of genetic mutation to cyber infrastructure with the aim of improving resiliency against active cyber threats e.g. Denial of Service (DoS) attacks. They propose to change different parameters of the network (e.g. Routes or IP addresses) proactively to avoid links under congestion and to avoid spreading of malware [10, 16, 23]. This domain is relatively new (and less explored) as compared to other bio-inspired cyber-security domains. Therefore, decentralized and more efficient algorithm are required to fill the gap.

The fourth and the most important area which has not been explored by the researchers so far, to its fullest potential, is the study of natural phenomenon of Cell and Gene regulation via signal transduction [22]. *Signal Transduction* is a mechanism in which (observed) exterior signals from a cell are transmitted into its interior against which numerous autonomous entities, i.e., genes regulate each other to generate an appropriate response and maintain homeostasis (by maintaining the optimal values of different parameters, e.g., blood pressure, body temperature, and sugar level). The first and the only practical contribution in this domain was proposed by Dressler [9]. The author proposed a refined and practical model for self-organization in a network facility (i.e., load management during packet inspection in intrusion detection systems) based on the concept of cell regulation [9]. The motivation behind the choice was to embed self-organization in a distributed detection system to regulate the amount of traffic rate between probes and detection system in variable situations, and in order to save the detection unit from becoming a potential target.

In the modern cyber networks, the usability of applications or services running on the end hosts are very important for an enterprize and hence cannot be ignored. Although the approach proposed by Falko Dressler, presents an auto-regulatory architecture for distributed IDS, the model does not incorporate the notion of usability associated with an end host and risk affiliated with a flow. The reactive strategy of the mentioned approach completely relies on the maximum throughput that a detection unit can handle without any regard to the usability of end host, reachability requirement, and potential risk imposed by a certain flow.

3   Biological Regulatory Networks (BRNs)

Every functionality in the human body and evolution of the morphological features is highly influenced or controlled at molecular level [6]. Genes and proteins are the main ingredient of this controlling mechanism, which cooperate together in a programmed manner to perform multiple tasks in an organism. Genes are the informative subunits of the DNA and they decode instructions in the form of proteins. Some proteins have the function of regulating the expression of genes by turning them on or off. This process of interaction, between genes and protein regulatory elements, establishes a BRN.

BRNs are very unique in their functionality as they have tendency to operate in adverse conditions under extreme threats without any central control or external monitoring. The main strength and capability of exceptional operability come from the structure of interactions, which is a feedback control mechanism. It is only through feedback loops in a BRN that imposes a controlled mechanism in order to maintain an optimal concentration of proteins in a cell [29]. Such feedback loops give rise to the phenomenon of genetic oscillations, which play a main role in the activity of maintaining the cascade of internal biochemical reactions with the extracellular environment. Molecular alterations in the performance of such behavioral rhythms can lead to the severe pathological diseases, e.g., cancer. This biological phenomenon can be summarized in a simple way: the dynamics of the living system is controlled by the BRNs, and at any given time a BRN of a living organism should optimize the cell behavior by maintaining the concentrations of proteins to make it survive in its (often abnormal) environmental conditions.

The behavior of a single entity in a BRN can be classified into the sequence of three different functions which are repeated infinitely often [25]: Sensing (signals from the neighboring entities), Actuating (changing internal state) and Signaling/influencing (firing/triggering an output signal) neighboring entities. A biological entity (gene/cell) in a BRN can either influence its neighboring entities positively or negatively. The process of positively (conversely negatively) influencing others is referred as *Activation* (conversely *Inhibition*). The influence phase of this natural process (activating/inhibiting neighboring entities), forms a feedback loop which is very fundamental to the control mechanism in BRNs for maintaining the optimal value of different parameters. As a first step towards creating bio-inspired resilient architecture, we intend to understand how this feedback notion works and can help us accomplishing our objectives. In the next section, we present a real-life example of a BRN, which is responsible for respiratory mechanism in the human body and demonstrate how its malfunctioning can lead to a severe lung disease.

3.1   BRN of the Cystic Fibrosis (Pseudomonas Aeruginosa)

Cystic fibrosis is a life threatening genetic disease that primarily effects the lungs and digestive systems [24]. The main cause of the respiratory deficiency in patients of cystic fibrosis is mucus production. The regulatory network which controls the mechanism of mucus production is shown in Fig. 1. $AlgU$ (x) is the main regulator of mucus production and it favors its own production while another gene inhibits it. The regulatory network of mucus production can be analyzed using different approaches (e.g. Linear Hybrid Automata, coupled Differential Equation, Regulatory Network Transition Systems, or Regulatory Graphs)

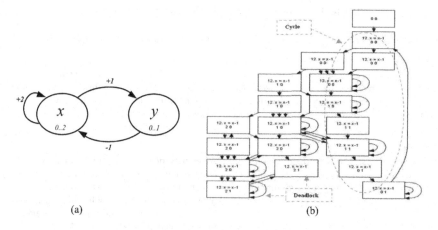

(a)   (b)

Fig. 1. (a) BRN of Pseudomonas aeruginosa, (b) Corresponding state space

[24], but for the convenience and simplicity we represent it by a regulatory graph in Fig. 1(a), where x represents gene $AlgU$ (or its protein) and y represents the inhibitor protein of $AlgU$. The concentration of x and y is defined over the qualitative levels. A positive sign (+) represents that x is the promotor of y or positively influences y when its (x's) concentration reaches level-1. As a result of positive influence (activation), y's concentration starts increasing. Once y's concentration reaches level 1, it inhibits x; as a result the concentration of x reduces towards the minimum (through a feedback mechanism). The regulatory interaction "+2" means that x becomes its own promoter/activator once its concentration reaches level 2. The concentration levels of each biological entity are represented qualitatively. Level '0' corresponds to the situation when the concentration for the protein of a certain gene/biological entity is absent. In the same manner, the higher levels, i.e., '1' or '2' refer to certain amount of concentration. By carefully analyzing the regulatory graph of the Cystic Fibrosis, Rauf et al. [25] show that it can govern two types of behaviors, oscillatory and deadlock. Oscillatory is considered as normal whereas the deadlock condition which is a hold and wait event sequence can be referred to as a malicious behavior. If the inhibitory entity (y) activates before a certain level (before which $AlgU$ favors its own production) then the system will not lead towards the malicious disease and will remain in progressive cycles, by maintaining the optimal values of the concentrations. Figure 1(b) shows the two possible behaviors of the (Pseudomonas aeruginosa) BRN, achieved by the concurrent model checking. Although there are two possible behaviors, the chances for malfunctioning of this BRN are very low as the probability of someone being infected by this disease is ≈0.00001. This is only because of the nature imposed feedback control mechanism that regulatory graph always avoids malicious behavior by remaining in progressive cycles/oscillations. To avoid deadlock, disease state or malicious state, there must be a realization of risk and there should be a feedback mechanism through

Table 1. Analogy between BRNs and cyber networks

Characteristics of biological entity	Characteristics of cyber entity
Sensing	
An entity receives signals from all the neighboring entities as a result it becomes aware of the current state of its neighboring entities	A security device can collect required information from its neighboring devices i.e. value of assets, desired reachability requirements, and risk evaluation against different policy rules
Actuation	
Biological entity increases/decreases its concentration under excitatory/inhibitory interaction from its neighboring entities	A security device can (autonomously) add/remove set of rules under the influence of certain activities/signals from its neighboring entities, which will result in increase/decrease of the attack coverage of a firewall/security device
Reaction	
Influencing neighboring entities through excitation or inhibition. However, nature of influence in biological systems is always static leading to disease states	A security device can also react against its actuation/updated (local) goals to influence/effect its neighboring entities, after evaluation of threat impact, risk payoff, or benefits associated with the assets, which it is responsible for the security and performance of the system

which y can be aware of the threats and can inhibit x. This shows that the understanding of theories and principles behind this self-organization through feedback mechanism is crucial.

4   BRN-Inspired Cyber-Security Architecture Mapping

To proceed further with the idea of integrating auto-resiliency characteristics of the biological systems in current cyber architecture, there must exist some analogy and mapping of actions among fundamental entities of both domains. Table 1 shows our effort towards mapping of BRNs dynamics to cyber networks.

4.1   Architecture of the Bio-inspired Firewall Regulatory Networks (FRNs)

We propose the idea of an architecture, through which security devices can regulate each other via inhibitory/excitatory orders to optimize the global objective. Figure 2 gives the realization of the proposed architecture, in which each security device has its own sensors and actuators for self-awareness and interaction with the neighboring elements. The most important part is the decision engine, which receives information from the sensors and decides which actions to take through actuators (according to global interest).

Nature of the Interaction can be of two types in terms of cyber networks. **Inhibition** order (from one device to another), which is driven by the risk associated to the assets for which a device is responsible and **Activation/Excitation** order, which is driven by pay-off associated with the reachability. In working principle of BRNs, nature of an interaction of a certain biological entity (gene) with its neighboring entities always remain static and does not change over time. The static (nature of) interaction leads towards disease states and malicious functionality of organs. We aim to avoid rather than eliminate the causes which may lead a system to bad/disease states.

Therefore, we propose that at any given time elements (security devices) of the network should synthesize a set of regulatory interactions (among them) which always leads towards progressive cycles (auto-regulation of the system) rather than deadlock/malicious states (where the risk is always high). In the next section we classify feedback mechanisms, which is fundamental to synthesize regulatory interactions at a given time; through which system can always remain in progressive cycles (self-organize or reconfigure itself if there is any perturbation in the external environment).

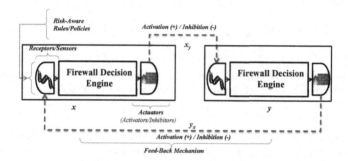

Fig. 2. Bio-inspired architecture of firewall regulatory networks for self organization/automated risk mitigation

4.2 Notion of the Feedback and Current Cyber Infrastructures

René Thomas and d'Ari [29] mathematically proved using Linear Stability Analysis that: *"A feedback loop is positive if it contains even number of negative interactions (Fig. 3(a)) and negative if it contains odd number of negative interactions (Fig. 3(b))"*. When a system only contains negative feedback (conversely positive), which means it only has odd number of negatively regulated interactions, it tends to oscillate around a certain optimal value. The oscillatory behavior can also be referred to as a cyclic behavior, and such mechanism in biological sciences is called "homeostasis". If synthesis/evolution of a certain parameter starts, it triggers the evolution of the same parameter in the following linked elements (through positive regulations), unless a negatively regulated entity is activated, from that point onward (the negatively regulated entity), suppresses the evolution of the same parameter in the following entities. This causes suppression in the synthesis of that entity by pushing it to the ground state where it is unable to regulate the neighboring entities. Hence the decay effect reaches to the original entity/element. This happens periodically and corresponding behavior is referred to as oscillatory behavior. Conversely, the systems having positive feedback loops tends to end up in (unique or multi) stable states/deadlocks. Stable states are those states in which the value of a certain parameter for all entities becomes stable and it remains the same, which can also be considered as a deadlock state, from where no progress becomes possible.

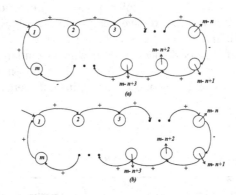

Fig. 3. Classification of feedback control mechanisms; (a) positive feedback loop, (b) negative feedback loop; where m and n represent entity indexing

In the context of current risk aware cyber infrastructures, deadlocks or stable-steady states can be viewed as states where the overall risk for an organization is above a certain bearable threshold. In the next section we describe our proposed framework in detail for synthesizing the correct feedback control models of Firewall Regulatory Networks (FRNs).

5   The Proposed Framework

This section describes our proposed BRN-inspired security architecture. Since, there are a variety of security devices for securing cyber networks, here we consider one particular type of security device, i.e., the network firewall (FW). A typical corporate network can use several firewalls to segregate the network according to the organizational needs, e.g., external, internal and demilitarized zone (DMZ). We propose a revolutionary new paradigm whereby all these FWs interact and form a network that we call FRNs. This model can be extended to any other network security devices in the future, such as, IRNs for intrusion detection regulatory networks where the host and the network IDS sensors interact.

5.1   FRN Synthesizer

Figure 4 sketches the details of our proposed framework, which uses FRN topology, connectivity requirements, and asset demands/responses as an input. The first step towards achieving a correct feedback control mechanism (for self-organization) is to synthesis the set of interactions at a certain point in time among all security devices (to resolve conflicting issues), so that bad/malicious states can be avoided. Describing the formal notion of cyber demand/response (between security firewalls) and the correct control logic are the fundamental ingredients of our proposed framework. For this purpose in the forthcoming section, we formalize cyber demand/response (regulatory interactions) features and control logic as Constraint Satisfaction Problem (CSP) [8,26].

As a next step in FRN synthesizing phase, we pass the formal model to the SMT solver z3 [7] to synthesize the FRN with the correct feedback control. Finally, we formally model the synthesized instance of the FRN, in concurrent model checking tool SPIN using PROcess MEta LAnguage (PROMELA) [15] to verify whether the synthesized instance achieves self-regulation or not. The details about each component of the framework are discussed in the forthcoming subsections.

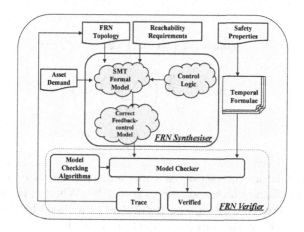

Fig. 4. Proposed framework

Formalization of the Request/Response and Reachability Requirements. The following set of constraints represents the nature of interactions (among two entities in a network) in the first order logic.

$$Inf_{i,j}^{(+,des)} : A_{i,des}^{req} \bigwedge (reach_{j,des} \bigvee j == des)$$

$$Inf_{i,j}^{(-,des)} : A_{i,des}^{resp} \bigwedge (reach_{j,des} \bigvee j == des)$$

$$\forall_{i,j} \left(Inf_{i,j}^{(+,des)} \bigvee Inf_{i,j}^{(-,des)} \right) \mapsto \mathcal{E}; \text{ where } i, j, \& des \in F$$

Where $Inf_{i,j}$ represents the nature of the influence entity i has on entity j, as a result of accessability request ($A_{i,des}^{req}$ between i and des) or as a result of response ($A_{i,des}^{resp}$), and F is a set of security devices in a regulatory network. The nature of influence can be $+$ or $-$, depending upon if it is a request (req) or a response ($resp$). Note that we make a basic assumption here, i.e., positive influence ($+$) is derived by an accessability demand ($A_{i,des}^{req}$) from a source (i) to a destination (des) and negative sign ($-$) is influenced by the risk affiliated with response ($A_{i,des}^{resp}$) of accessability demand. Therefore, influence of an entity (i) on a neighboring entity (j) against an accessability demand/response can be calculated by evaluation of ($Inf_{i,j}^{(+,des)} \bigvee Inf_{i,j}^{(-,des)}$). In any case, all the edges across the network ($e \subseteq \mathcal{E}$) should be assigned a value against any request/response between i and des. These constraints must be evaluated against every accessability request from a source to a destination (des) over the whole FRN, to figure out the nature of the influence among security devices.

Formalization of the Feedback Control Logic. In the following constraints, we formalize the correct feed-back control logic, which must be consistent throughout the regulatory network.

$$func : \mathbb{N} \longmapsto \mathbb{B}$$

$$func(e_i) : \left(\sum_{i:1}^{n} e_i \right) \%2$$

$$func(e_i^l) : \bigwedge_{l=1}^{m} \left(\left(\sum_{i:1}^{n} e_i^l \right) \%2 \right)$$

$$\forall_l \exists_i \left(func(e_i) \neq 0 \right) \longmapsto \mathbb{B}$$

We describe regulatory interaction as a mathematical function ($func$) over the set of natural numbers, which evaluates to true/false. Where e_i is the set of interactions (edges) between entities/security devices in a closed feedback loop, and there can be n such interactions (in a closed feedback loop), i.e., e_i: {e_1, e_2,..., e_n}. We represent $+ve$ (conversely $-ve$) interaction as a number 0 (conversely 1) due to its positive parity. Therefore, the mathematical function representing the regulatory interactions (in a closed feedback loop) according to the formalization is the sum of all interactions in a closed feedback loop with modulo 2. Finally there can be m multiple feedback loops in a system or a security device/component may be involved in multiple feedback mechanisms, thus, to achieve global objective we quantify over all closed feedback loops to find a satisfiable instance of the model. If there exists a satisfiable instance, we get an answer as true along with the configuration of the instance.

The above mentioned formalizations together provide us with the satisfiable instance of the system/model, which has an odd number of -ve interactions. As mathematically proved by René Thomas and Richard d'Ari, that an even number of -ve interactions leads system to a deadlock or malicious state, our formalization tends to avoid the instances of the system/model which may lead the system towards malicious behavior or which may not have tendency to optimally regulate the value of risk (globally) in a network [29].

5.2 FRN Verifier

To prove the correctness of the proposed formalism, we synthesize FRNs model based on the real-life examples, and verify it using model checking to determine whether the synthesized models are able to reconfigure themselves under any external perturbation. If the control logic is correctly integrated the corresponding models should be able to recover from the high risk states. As measuring quantitative risk is not the focus of this research, we affiliate qualitative risk levels to each entity in the FRN (as a firewall is represented as a state machine, which can evolve over these levels which in reality represents the risk imposed by the active policy in a device, and can be calculated using any risk assessment metric). We allow risk levels to evolve over the qualitative values, to observe if the resultant state space of the FRN contains any deadlock or not. At the end, we update our topology, as user accessability demands change over time, which might lead the previously synthesized model to a deadlock. Therefore, whenever additional accessability demands arrive, new model must be synthesized. For the verification purposes, we present an abstract formal representation of a firewall/security device, regulatory interactions, and the parameters which govern the state of a security device.

In our abstraction, the state of a firewall is defined over qualitative levels. This means that a firewall is an entity which can assume any qualitative value from a given set. The firewalls can have an impact on each other and such impact is modeled as regulatory interactions. As utility of any active subset of firewall rules can be calculated and thresholds on regulatory interactions can also be assigned, therefore, our abstraction is practical and aligned with the real-life practice. In the following sections we discuss the modeling elements of the **Verifier** one by one in details.

Discrete Model of a Firewall. In this section we present the formal model of an entity in a regulatory network. A regulating entity (e.g. firewall) is defined as an automaton. It receives an input from interacting neighbors, changes its internal state in response to it, and produces an appropriate output, depending on threshold level (θ_{ij}), where i, j are the interacting entities.

Formally, a set of firewalls F can be expressed as a set of interacting automata and a firewall may assume any positive value in a range.

$$F = \{f_1, f_2.., f_m\};$$
$$f_k = \{0, \ldots, n_k\}; \; where\, k \in \{1, \ldots, m\}$$

There can be m firewalls in a network and any firewall f_k can have any possible discrete qualitative levels. The possible states for the network $\mathcal{F}$ can then be defined as the cartesian product:

$$\mathcal{F} = f_1 \times f_2 \times f_3 \times \ldots \times f_m$$

Regulatory Interactions Modeling. An excitatory (resp. inhibitory) interaction $(f_1 \xrightarrow{+} f_2)$ (resp. $f_1 \xrightarrow{-} f_2$) is active when usability demand to access certain area is equal to or above a specific threshold level θ. Conversely, inhibitory interaction $(f_1 \xrightarrow{-} f_2)$ is active or triggered when the risk imposed on a certain firewall is equal or above a certain threshold. We also associate a threshold (θ_{12}) to each interaction from $(f_1 \xrightarrow{\theta_{12}} f_2)$.Where $\theta_{12} \in \{1, ..n\}$. f_1 is called the activator of f_2, if $f_1 \geq \theta_{12}$ (resp. $f_1 < \theta_{12}$) for the excitatory interaction (respectively inhibitory interaction).

Modeling of Parameters. At any time instant the state of a firewall depends only on its set of attractors. Attractors are the other entities (firewalls) which can directly influence a firewall via inhibitory or excitatory interactions. We represent the set of attractors of an entity as $w(f_i^\alpha)$, where $\alpha \subset \{1, ..., m\}$. The residual effect of $w(f_i^\alpha)$ on the evolution of f_i^α can be given by the logical parameter.

$$\mathcal{K}(w(f_i^\alpha)) \in \{0, \ldots, n_\alpha\}$$

The logical parameter corresponds to the level towards which a firewall evolves:

1. if $f_i^\alpha < \mathcal{K}(w(f_i^\alpha))$ then f_i^α is increasing
2. if $f_i^\alpha > \mathcal{K}(w(f_i^\alpha))$ then f_i^α is decreasing
3. if $f_i^\alpha = \mathcal{K}(w(f_i^\alpha))$ then f_i^α is stable

The above mentioned behavior of evolution of a firewall over qualitative levels, as a response of interactions with neighboring entities is modeled as *Resource_allocation* process in SPIN model checker.

The Complete Model of the FRN. The main components of a FRN are:

- m security devices (e.g. firewalls) are modeled as m processes f_i where $i = \{1, 2, \ldots, m\}$ having their own identical thresholds for either usability or risk.
- Process *Resource_ allocation* which changes resources of f_i as its internal states are changed as a result of interactions, keeping in view its attractors $\mathcal{K}(w(f_i^\alpha))$.
- Process *Observer* which ensures at every step of the computation that f_i remains within the bound $\{0, ..., n\}$.

After modeling of all components of the system, namely f_i, *Resource_allocation*, and *Observer*, we make parallel composition of all the components and allow the system to evolve.

$$FRN : \; f_i \; || \; Resource\_allocation \; || \; Observer$$

6   Evaluation

For the synthesis of the regulatory interaction between security devices for control mechanism (FRN), we use Z3 SMT solver [7]. For verifying the correctness of the generated instances of FRN, we use PROMELA as modeling formalism and model checking tool SPIN [15]. All the experiments were conducted on Core i7 machine with 3.4 GHz processor, and 16 GB memory.

6.1   Case Studies

Regulation of Risk Between Two Security Devices. We consider a real-life scenario in which few assets in local area network are protected by a screening firewall and a specialized firewall. Figure 5(a) gives the description of the case study. Customer service department in the Demilitarized Zone (DMZ) is protected by an immediate screening firewall (Fw_1), whereas the more important (A_2, A_4) and critical (A_1, and A_3) assets are protected by a specialized firewall (Fw_2). Due to the desire for attracting customers/visitors, web servers have influence on screening firewall to increase its "allow" space. Lets assume that to facilitate the customers, Fw_1 (managing the customer service department) sends a service accessability request (Skype) for all assets behind Fw_2 (managing the engineering department). Since only Fw_2 has a complete visibility of the assets under its control, it is aware of a known Skype's elevation of privileges vulnerability (CVE-2017-11786), which resides at A_2 (a non-critical asset). Fw_2 evaluates the impact of the request and realizes that it imposes a threat (with CVSS v3.0 base score of 88%) to its neighboring critical assets (A_1, A_3) as well. As a consequence, it must inhibit Fw_1, if the risk imposed is above the pre-specified threshold. In Fig. 5, X represents Fw_1 and Y represents Fw_2.

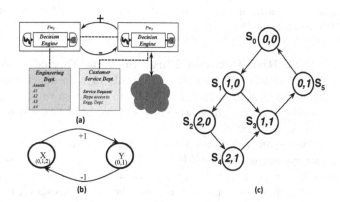

Fig. 5. (a) Case study: two firewalls securing assets, (b) Synthesized FRN model, (c) Corresponding State Space

Now to resolve this issue, we need reconciliation, and before that we need a control mechanism, representing: *"who can influence whom and in what sense"*. Using our formalism to synthesize the regulatory interaction, we obtain the model in Fig. 5(b). We associate qualitative numbers as threshold for verification. The resultant model states: when the demand of customer services X increases and reaches up to a qualitative level 1, it imposes a threat on the firewall Y. As a result of this situation the risk on Y starts increasing. When the risk of Y reaches an unbearable threshold, it transmits an inhibition order to X. The inhibition order informs X to reduce its allow space (or to only activate the set of flows/policies for which the residual risk is below the qualitative level 1), until the threat imposed on Y is reduced to the minimum. To prove correctness of the proposed approach we model the derived instance of the system in SPIN model checker. As both firewalls are concurrently evolving or dynamically changing entities which influence each other, therefore, PROMELA is the best suited formalism to analyze the behavior of such concurrent model.

We verify correctness of our postulate about automated risk mitigation using correct feedback control mechanism via self-regulation property written in Linear Temporal Logic (LTL):

$$(x = 0, y = 0) \implies \mathbf{X}(\neg(x = 0, y = 0)) \wedge$$
$$\mathbf{G}(\mathbf{F}(x = 0, y = 0))$$

The property states that: *if the system starts from a normal state where $x = 0$ & $y = 0$, and the next state is not $x = 0$ & $y = 0$, then along the path of evolution, in future the system eventually goes back to the ground or normal state $x = 0$ & $y = 0$.*

Figure 5(c) illustrates the existence of self-regulation in the state space of the model with two regulatory firewalls, it also shows how system regulates itself or recovers once it reaches malicious/bad state ($x = 2$ & $y = 1$), where risk is maximum or above a normal value. The state space generated in the example is from the SPIN model checker.

Regulation of the Risk Between Three Security Devices. We present another case study which contains three firewalls in a feedback mechanism. Figure 6 gives the description of the case study. In this case we consider a bad state to be a situation where (qualitative levels of) residual risk always remains maximum ($X = 2$ & $Y = 1$ & $Z = 2$).

To verify if our proposed method works with the extended case study we again model our system using PROMELA [15]. The state space of the system (Fig. 7) shows that the system never gets stuck in a malicious state. Whenever it encounters a situation where the risk is above bearable threshold, it regulates (recovers) itself and remains in a progressive cycle. Careful analysis shows that there are multiple ways to avoid the high-risk state. The best possibility is when firewall Z inhibits firewall X before it (X) reaches a configuration where its risk becomes maximum, as a result the system switches from (1,1,1) to (0,1,1). In the worst case scenario, the system reaches the state (2,1,2), due to the delayed

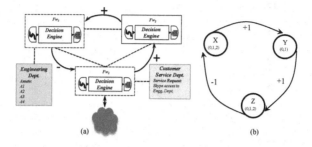

Fig. 6. (a) Case study: three firewalls securing assets, (b) Synthesized FRN model

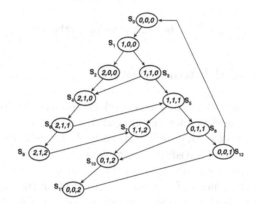

Fig. 7. State space of the FRN containing three regulatory firewalls

propagation of inhibition order from Z to X. Although it reaches malicious state, it recovers from it and goes back to the normal states after series of alterations in its configuration $((2,1,2) \rightarrow (0,1,2) \rightarrow (0,0,2) \rightarrow (0,0,1) \rightarrow (0,0,0))$.

Overhead of the SMT Formalization and Formal Verification. We also perform overhead analysis of the SMT formalization by selecting different number of entities in a feedback loop (to synthesize regulatory interaction). Figure 8 illustrates that our formalization is capable of synthesizing a set of regulatory interactions for a large number of devices in milliseconds. As we increase the number of entities the required time increases linearly which makes this approach feasible and practical for the real-life implementation.

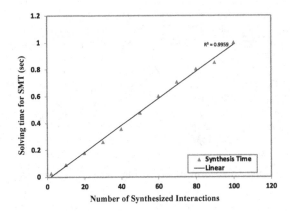

Fig. 8. Overhead analysis of synthesizer

7   Conclusion and Future Work

In this paper, we present a bio-inspired resilient architecture for cyber-security. We propose a novel framework for automated risk mitigation by embedding self-organizing features in the current infrastructure. By formalizing feedback control mechanism as a constraint satisfaction problem in SMT we allow for an automatic synthesis of cyber-regulatory networks which can reconfigure themselves, thereby, eliminating the need of manual reconfiguration by the administrators. To prove the correctness our proposed approach, we formally model two real-life scenarios and analyze their self-organizing behavior using model checking tool. The results show that the proposed architecture allows cyber network to be self-organizing, and dynamically adaptable to variable conditions. The low overhead for synthesizing correct feedback control makes our approach more practical for real-life implementation. In the future we plan to deploy this architecture on the medium scale networks and determine its resiliency against different scenarios. We also aim to explore cooperative game theory, to embed notion of conflict resolution in this architecture to deal with conflicting scenarios, in which a device might not be interested to cooperate with the other devices.

References

1. International Standards Organization ISO/IEC 27005: 2008. Information technology-security techniques-information security risk management. International Standards Organization, Geneva, Switzerland (2008)
2. Aickelin, U., Bentley, P.J., Cayzer, S., Kim, J., McLeod, J.: Danger theory: the link between AIS and IDS. CoRR, abs/0803.1997 (2008)
3. Bonabeau, E., Dorigo, M., Theraulaz, G.: Swarm Intelligence: From Natural to Artificial Systems. Oxford University Press Inc., New York (1999)
4. de Castro, L.N.: Artificial Immune Systems: A New Computational Intelligence Approach. Springer, London (2002)

5. de Castro, L.N., Von Zuben, F.J.: The clonal selection algorithm with engineering applications. In: GECCO - Workshop Proceedings, pp. 36–37. Morgan Kaufman (2002)

6. Davidson, E.H., Erwin, D.H.: Gene regulatory networks and the evolution of animal body plans. Science **311**(5762), 796–800 (2006)

7. de Moura, L., Bjørner, N.: Z3: an efficient SMT solver. In: Ramakrishnan, C.R., Rehof, J. (eds.) TACAS 2008. LNCS, vol. 4963, pp. 337–340. Springer, Heidelberg (2008). https://doi.org/10.1007/978-3-540-78800-3_24

8. Dechter, R.: Constraint Processing. Morgan Kaufmann Publishers Inc., San Francisco (2003)

9. Dressler, F.: Self-organized network security facilities based on bio-inspired promoters and inhibitors. In: Dressler, F., Carreras, I. (eds.) Advances in Biologically Inspired Information Systems. Studies in Computational Intelligence, pp. 81–98. Springer, Heidelberg (2007). https://doi.org/10.1007/978-3-540-72693-7_5

10. Duan, Q., Al-Shaer, E., Jafarian, H.: Efficient random route mutation considering flow and network constraints. In: 2013 IEEE Conference on Communications and Network Security (CNS), pp. 260–268, October 2013

11. Farmer, J.D., Packard, N.H., Perelson, A.S.: The immune system, adaptation, and machine learning. Physica D **22**, 187–204 (1986). Proceedings of the Fifth Annual International Conference

12. Fink, G.A., Haack, J.N., McKinnon, A.D., Fulp, E.W.: Defense on the move: ant-based cyber defense. IEEE Secur. Priv. **12**(2), 36–43 (2014)

13. Forrest, S., Perelson, A.S., Allen, L., Cherukuri, R.: Self-nonself discrimination in a computer. In: Proceedings of 1994 IEEE Computer Society Symposium on Research in Security and Privacy, pp. 202–212, May 1994

14. Haack, J.N., Fink, G.A., Maiden, W.M., McKinnon, A.D., Templeton, S.J., Fulp, E.W.: Ant-based cyber security. In: 2011 Eighth International Conference on Information Technology: New Generations (ITNG), pp. 918–926, April 2011

15. Holzmann, G.J.: The SPIN Model Checker: Primer and Reference Manual. Addison-Wesley Professional, Boston (2003)

16. Jafarian, J.H., Al-Shaer, E., Duan, Q.: Openflow random host mutation: transparent moving target defense using software defined networking. In: Proceedings of the First Workshop on Hot Topics in Software Defined Networks, HotSDN 2012, pp. 127–132. ACM (2012)

17. Jinquan, Z., Xiaojie, L., Tao, L., Caiming, L., Lingxi, P., Feixian, S.: A self-adaptive negative selection algorithm used for anomaly detection. Prog. Nat. Sci. **19**(2), 261–266 (2009)

18. Li, G.Y., Guo, T.: Receptor editing-inspired negative selection algorithm. In: 2010 International Conference on Machine Learning and Cybernetics (ICMLC), vol. 6, pp. 3117–3122, July 2010

19. Liu, Z., Kwiatkowska, M., Constantinou, C.: A swarm intelligence routing algorithm for manets. In Proceedings of the 3rd IASTED International Conference on Communications, Internet and Information Technology (CIIT 2004), p. 1. ACTA Press (2004)

20. Modi, P.J., Shen, W.M., Tambe, M., Yokoo, M.: Adopt: asynchronous distributed constraint optimization with quality guarantees. Artif. Intell. **161**(1), 149–180 (2005)

21. Muraleedharan, R., Osadciw, L.A.: An intrusion detection framework for sensor networks using honeypot and swarm intelligence. In: 6th Annual International Mobile and Ubiquitous Systems: Networking Services, MobiQuitous 2009, pp. 1–2, July 2009

22. Rauf, U.: A taxonomy of bio-inspired cyber security approaches: existing techniques and future directions. Arab. J. Sci. Eng. **43**, 6693–6708 (2018)

23. Rauf, U., Gillani, F., Al-Shaer, E., Halappanavar, M., Chatterjee, S., Oehmen, C.: Formal approach for resilient reachability based on end-system route agility. In: Proceedings of the 2016 ACM Workshop on Moving Target Defense (MTD), pp. 117–127 (2016)

24. Rauf, U., Sameen, S., Cerone, A.: Formal analysis of oscillatory behaviors in biological regulatory networks: an alternative approach. Electron. Notes Theoret. Comput. Sci. **299**, 85–100 (2013)

25. Rauf, U., Siddique, U., Ahmad, J., Niazi, U.: Formal modeling and analysis of biological regulatory networks using spin. In: 2011 IEEE International Conference on Bioinformatics and Biomedicine (BIBM), pp. 304–308, November 2011

26. Rossi, F., van Beek, P., Walsh, T.: Handbook of Constraint Programming (Foundations of Artificial Intelligence). Elsevier Science Inc., New York (2006)

27. Sellami, K., Chelouah, R., Sellami, L., Ahmed Nacer, M.: Intrusion detection based on swarm intelligence using mobile agent. In: International Conference on Swarm Intelligence, June 2011

28. NIST SP800-30. Risk Management Guide for Information Technology Systems. National Institute of Standards and Technology, USA (2002)

29. Thomas, L.C., d'Ari, R.: Biological Feedback. CRC Press, Boca Raton (1990)

30. Zeng, J., Liu, X., Li, T., Li, G., Li, H., Zeng, J.: A novel intrusion detection approach learned from the change of antibody concentration in biological immune response. Appl. Intell. **35**(1), 41–62 (2011)

Space Partitioning and Maze Solving by Bacteria

Ayyappasamy Sudalaiyadum Perumal[1], Monalisha Nayak[1],
Viola Tokárová[1,2], Ondřej Kašpar[1,2], and Dan V. Nicolau[1(✉)]

[1] Department of Bioengineering, Faculty of Engineering, McGill University,
Montreal, QC H3A 0C3, Canada
dan.nicolau@mcgill.ca
[2] University of Chemistry and Technology, Prague,
Prague 166 28, Czech Republic

Abstract. Many bacteria dwell in micro-habitats, e.g., animal or plant tissues, waste matter, and soil. Consequently, bacterial space searching and partitioning is critical to their survival. However, the vast majority of studies regarding the motility of bacteria have been performed in open environments. To fill this gap in knowledge, we studied the behaviour of *E. coli K12-wt* in microfluidic channels with sub-10 μm dimensions, which present two types of geometries, namely a diamond-like network and a maze. The velocity, average time spent, and distance required to exit the networks, have been calculated to assess the intelligent-like behaviour of bacteria.

Keywords: Bacterial motility · Microfluidics · Maze

1 Introduction

Solving mazes is a nontrivial exercise that has been used for testing the intelligent-like behaviour of many organisms, e.g., ants [1], bees [2], rats [3], octopi [4, 5] and humans [6], as well as robots [7, 8] and rat-cyborgs [7]. Interestingly, even simple organisms, such as slime mold and fungi [9–11] possess complex and efficient biological algorithms employed in space searching and partitioning. However, despite their ubiquity, and despite studies regarding the motility characteristics of *E. coli* and its variants in straight microfluidics channels and directional preferences [12], or in simple geometries [13, 14], bacteria have not been the subject of an assessment of their intelligent-like behaviour via the classical exploration of mazes. Furthermore, earlier studies use chemotactic/attractant-based solution seeking studies in a maze [8–10], but chemotaxis is inferring with bacterial innate space searching capabilities, which are essential in nutrient poor environments. To fill this gap in knowledge, we studied the maze solving abilities of a common lab host, *Escherichia. coli K12*-wt, using simple microfluidics networks depleted of chemotactic clues.

© ICST Institute for Computer Sciences, Social Informatics and Telecommunications Engineering 2019
Published by Springer Nature Switzerland AG 2019. All Rights Reserved
A. Compagnoni et al. (Eds.): BICT 2019, LNICST 289, pp. 175–180, 2019.
https://doi.org/10.1007/978-3-030-24202-2_13

2 Materials and Methods

Poly di(methyl) siloxane, PDMS, was used to fabricate the confining network, via soft lithography replication using a silicon master, in turn fabricated using optical lithography. The designs of the microfluidics networks comprise a simple, uniform diamond-shaped maze, and a more complex, non-uniform maze. The height of the channels is set up to 6 µm and the widths are 3 µm. The protocol for the fabrication of microfluidics networks has been described earlier [12].

The PDMS microfluidics networks were plasma-treated, to render them hydrophilic, then sealed on to the coverslips, pre-wetted in a Petri dish with LB medium, and stored at low temperature before use for experiments. For the bacterial motility experiments, the PDMS structures were explored by a log phase culture of fluorescently labelled *E. coli K12* (Plasmid, pmf-440mcherry, add gene plasmid #62550 received a gift from Prof. Michael Franklin's lab). The fluid environment is a nutrient-rich medium, i.e., Luria-Bertani (LB) broth. Furthermore, the observation times were short enough to precluded and discernable consumption of nutrients resulting in concentration gradients. This experimental methodology would provide answers not only to explore the shortest paths but also various paths that may contribute to efficient searching of the space in a geometrically complex microenvironment.

The movement of bacteria in microfluidics networks was studied using a fluorescence microscope (Olympus IX83) with a 40X objective. The frames were analysed by ImageJ open software with plug-ins, e.g., track mate and MtrackJ [15]. We used density maps generated from background subtracted image stacks. The motility was recorded as tracks of x-y coordinates, which were used for the calculation of several other physical parameters.

3 Results and Discussion

3.1 Diamond Networks

The diamond structures present to bacteria two types of routes of the shortest path from entry to exit, i.e., the outer boundary routes (Fig. 1A(ii)), of and the zig-zagged path (Fig. 1A(iii)). Because of its symmetrical nature, in the uniform maze (diamond structure), the shortest path is equivalent irrespective of the route that the bacteria take (provided the bacterial agents do not revisit any path). The heat patterns of the trajectories in Fig. 2A suggest that *E. coli* prefers the outer straight-line paths (as shown in Fig. 1A(ii)). Although the zigzagged inner paths were also explored, these instances were rare and they happened when the outer boundary channels were crowded.

3.2 Maze Structures

Similarly, to the study of the bacterial motility in the diamond networks, we calculated the performance parameters for non-uniform mazes. Contrary to the results in diamond structures, the inner paths were well explored, and the heat-map reached saturation. It can be observed (table (Fig. 2C)) that *E. coli* spent more time in the non-uniform maze

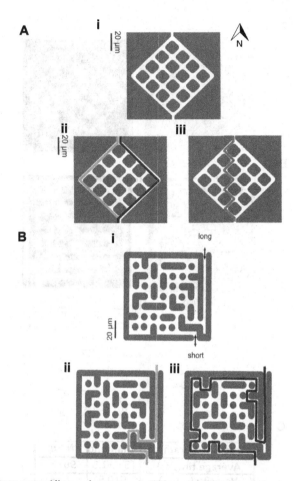

Fig. 1. A. Uniform maze (diamond structures): (i) layout of the diamond maze, (ii) and (iii) few different possible routes in the structure with the simplest (ii – green and red) and a zig-zag path (iii - pink). B. Non-uniform maze: (i) layout of the maze with two types of entry, short and long, (ii) and (iii) examples the shortest path possible (ii-green); and the longest path (iii – red). In both mazes, a bacterium could spend any range of time between seconds to hours, as the mazes are allowed for multiple solutions. Therefore, the paths presented here are only a few from possible paths. (Color figure online)

compared to the diamond structures, possibly due to the geometries like circular pillars, corners and barriers that deflected and diverted the bacteria visiting already explored paths.

Apart from the above-mentioned qualitative and empirical observations, we studied other motility characteristics, i.e., the preferences for long, and short points of the maze, and directional preferences. Interestingly, *E. coli K12* appears to use a 'wall-follower algorithm' which helps to solve the maze faster that would be otherwise achieved by a random exploration. This natural algorithm could be the result of the propensity of *E. coli* to swim closer to a surface ("wall accumulator" behaviour). This hypothesis

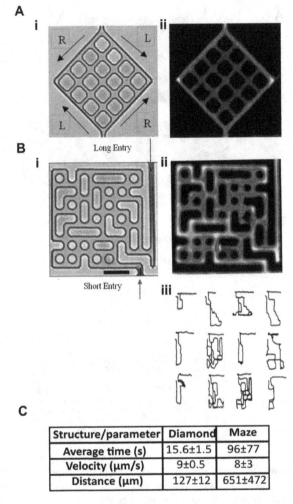

Fig. 2. A Motility of *E. coli K12*-wt in diamond structures - (i) Optical image of the structure under a microscope. (ii) Heat-maps/density map of bacterial trajectories in the diamond structure B. Motility of *E. coli K12*-wt in maze structure. (i) Optical image of the microfluidic device. (ii) Heat-maps/density maps of the bacterial trajectories in a non-uniform maze, (iii) few actual representative trajectories of the *E. coli K12* in the maze. C. Tabular column comparing three major parameters calculated from the motility of bacteria in the network.

needs to be verified by further experiments seeking to compare this natural algorithm for bacteria having "wall escaper" behaviour, e.g., *Magnetococcus marinus MC-1*.

Another layer of complexity is the emerging behaviour of space search algorithms resulting from extremely varied architectures of the flagellar system. Finally, in many cases bacteria vary the ratio and frequency of the run & tumble machinery in response to the environment parameters, which in turn will impact on the natural space search and partitioning natural algorithms. Future more comprehensive experiments involving various species of bacteria, each presenting different architectures, will advance in the understanding of the relationship between bacterial structure and resulting space search and partitioning algorithms.

4 Conclusion

In opposition to other studies, which studied bacterial motility in non-confining environments, the present work focused on the movement of bacteria in confining microfluidics structures, which could be conceived as a model of the microenvironments that they colonise. Furthermore, the present study operated in fluid conditions that suppress chemotaxis, which is one of the main driving forces of the directionality of bacterial movement. Consequently, the bacterial capacity to explore simple networks and more complex mazes, could be explored free of chemotaxis inference. This preliminary study suggests the further exploration of the motility patterns of other bacteria with different flagellar arrangements, and number, and exhibiting wall-escaping as well as wall accumulator character. Finally, this study suggests that bacteria can act as independent 'computational' agents solving mazes by employing space searching and partitioning natural algorithms.

Acknowledgements. The work presented here was financially supported by the Defence Advanced Research Projects Agency (DARPA) under Grant Agreement No. HR0011-16-2-0028.

References

1. Kohler, M., Wehner, R.: Idiosyncratic route-based memories in desert ants, Melophorus bagoti: how do they interact with path-integration vectors? Neurobiol. Learn. Mem. **83**(1), 1–12 (2005)
2. Jin, N., Landgraf, T., Klein, S., Menzel, R.: Walking bumblebees memorize panorama and local cues in a laboratory test of navigation. Anim. Behav. **97**, 13–23 (2014)
3. Cohen, J.S., Burkhart, P., Jones, N., Innis, N.K.: The effects of an intramaze cue search rule on rat's spatial working memory. Behav. Processes **22**(1–2), 73–88 (1990)
4. Lee, P.G.: Chemotaxis by Octopus maya Voss et Solis in a Y-maze. J. Exp. Mar. Biol. Ecol. **156**(1), 53–67 (1992)
5. Schoenfeld, R., Moenich, N., Mueller, F.J., Lehmann, W., Leplow, B.: Search strategies in a human water maze analogue analyzed with automatic classification methods. Behav. Brain Res. **208**(1), 169–177 (2010)
6. Slusny, S., Neruda, R., Vidnerova, P.: Comparison of behavior-based and planning techniques on the small robot maze exploration problem. Neural Netw. **23**(4), 560–567 (2010)
7. Yu, Y., et al.: Intelligence-augmented rat cyborgs in maze solving. PLoS ONE **11**(2), e0147754 (2016)
8. Asenova, E., Lin, H.Y., Fu, E., Nicolau, D.V., Nicolau, D.V.: Optimal fungal space searching algorithms. IEEE Trans. Nanobiosci. **15**, 613–618 (2016)
9. Nakagaki, T.: Smart behavior of true slime mold in a labyrinth. Res. Microbiol. **152**(9), 767–770 (2001)

10. Hanson, K.L., Nicolau, D.V., Filipponi, L., Wang, L., Lee, A.P., Nicolau, D.V.: Fungi use efficient algorithms for the exploration of microfluidic networks. Small **2**(10), 1212–1220 (2006)
11. Held, M., Lee, A.P., Edwards, C., Nicolau, D.V.: Microfluidics structures for probing the dynamic behaviour of filamentous fungi. Microelectron. Eng. **87**(5–8), 786–789 (2010)
12. Nayak, M., Perumal, A.S., Nicolau, D.V., van Delft, F.C.: Bacterial motility behaviour in sub-ten micron wide geometries. In: 2018 16th IEEE International New Circuits and Systems Conference (NEWCAS), pp. 382–384. IEEE (2018)
13. Männik, J., Driessen, R., Galajda, P., Keymer, J.E., Dekker, C.: Bacterial growth and motility in sub-micron constrictions. Proc. Natl. Acad. Sci. U. S. A. **106**(35), 14861–14866 (2009)
14. Libberton, B., Binz, M., Van Zalinge, H., Nicolau, D.V.: Efficiency of the flagellar propulsion of Escherichia coli in confined microfluidic geometries. Phys. Rev. E **99**, 012408 (2019)
15. Schneider, C.A., Rasband, W.S., Eliceiri, K.W.: NIH Image to ImageJ: 25 years of image analysis. Nat. Methods **9**(7), 671–675 (2012)

A Scalable Parallel Framework for Multicellular Communication in Bacterial Quorum Sensing

Satyaki Roy[1](✉), Mohammad Aminul Islam[2], Dipak Barua[2], and Sajal K. Das[1]

[1] Department of Computer Science,
Missouri University of Science and Technology, Rolla, MO, USA
{sr3k2,sdas}@mst.edu
[2] Department of Chemical and Biochemical Engineering,
Missouri University of Science and Technology, Rolla, MO, USA
{mixvc,baruad}@mst.edu

Abstract. Certain species of bacteria are capable of communicating through a mechanism called *Quorum Sensing (QS)* wherein they release and sense signaling molecules, called *autoinducers*, to and from the environment. Despite stochastic fluctuations, bacteria gradually achieve coordinated gene expression through QS, which in turn, help them better adapt to environmental adversities. Existing sequential approaches for modeling information exchange via QS for large cell populations are time and computational resource intensive, because the advancement in simulation time becomes significantly slower with the increase in molecular concentration. This paper presents a scalable parallel framework for modeling multicellular communication. Simulations show that our framework accurately models the molecular concentration dynamics of QS system, yielding better speed-up and CPU utilization than the existing sequential model that uses the exact Gillespie algorithm. We also discuss how our framework accommodates evolving population due to cell birth, death and heterogeneity due to noise. Furthermore, we analyze the performance of our framework vis-á-vis the effects of its data sampling interval and Gillespie computation time. Finally, we validate the scalability of the proposed framework by modeling population size up to 2000 bacterial cells.

Keywords: Autoinducer · Quorum Sensing · Gillespie ·
Multicellular system · Noise analysis · Population evolution ·
Scalability

1 Introduction

Large population of bacteria communicate with one another by releasing signaling molecules, called *autoinducers*, into the environment. Bacteria are also capable of sensing the environmental autoinducer concentration and regulating the expression of certain specific genes in a coordinated manner. This mechanism of communication and mutual regulation is called *Quorum Sensing (QS)*

S. Roy and M. A. Islam—Primary co-authors.

© ICST Institute for Computer Sciences, Social Informatics and Telecommunications Engineering 2019
Published by Springer Nature Switzerland AG 2019. All Rights Reserved
A. Compagnoni et al. (Eds.): BICT 2019, LNICST 289, pp. 181–194, 2019.
https://doi.org/10.1007/978-3-030-24202-2_14

[1]. Communication via QS has been observed in a wide range of bacteria species, such as marine bacteria (like *Vibrio fischeri*) and pathogenic bacteria [2,3].

Since each cell responds uniquely to its environment, any cellular regulation and signaling is prone to stochastic fluctuation. There have been attempts to study how population of bacteria achieve coordinated gene expression, despite such noise [4]. Sequential stochastic modeling of QS considers all reactions within the system, one reaction at a time. This makes modeling of large population of cells significantly more expensive, with respect to time and computational resources [2]. Thus, several works on parallel implementation of stochastic simulation of biological system have been proposed (though all of them do not pertain to QS alone). For example, a parallel software framework, using discrete agent-based simulation, has been proposed in [5]. It models the behavior of large cell population and updates molecular concentration using coupled Partial Differential Equations (PDE). A coarse-grained parallel approach is implemented in [6], to perform independent stochastic simulation. In [7], a C++ based stochastic and multi-scale simulation toolkit is proposed for chemically reacting system. The performance of Gillespie Stochastic Simulation is accelerated in [8] using Graphics Processing Units (GPUs). A parallel algorithm for off-lattice individual-based models of multicellular populations is presented in [9]. Gillespie's First Reaction is applied in [10] to present a stochastic simulation software framework for biochemical reaction networks. A graph-based model for parallel, distributed and portable applications is introduced in [11], and finally, a parallel algorithm is designed in [12], focusing on simulation of reaction-diffusion based system.

Let us consider an example of bacterial growth in rich media, where intercellular communication may be assumed to be negligible. Such a system can be implemented in parallel due to absence of significant dependency among the cells. However, in case of QS, cellular interaction via autoinducers plays a pivotal role in the coordinated behavior of the system. *An ideal parallel QS framework must therefore incorporate, both, modeling accuracy of molecular (especially autoinducer) concentration, as well as efficiency in terms of time and resource utilization.* Among the aforementioned literature, only [12] takes cellular interaction via environment into consideration. However, even that work neither discusses the inevitable trade-off between the parallelism and accuracy, nor the capability of accurately modeling population dynamics due to cell birth and death.

Contribution: In this paper we take the first step towards developing a scalable parallel framework for modeling biochemical network that meets both the requirements of accuracy and speed-up. We apply this framework to model QS in bacteria, where each cell is a process that exchanges messages with the master (or coordinator) process. It incorporates a simple approximation to maintain the uniformity of the environmental parameters. An initial version of this framework is discussed in [13]. Our simulation experiments show that this framework captures the dynamics of molecular concentration as accurately as the standard sequential QS model [2]. We analyze our system in light of how sampling interval affects the overall accuracy and variation of computation overhead due to varying concentration of molecules. We also discuss how this framework handles evolution due to cell birth and death. We show that our framework exhibits higher speed-up and more balanced CPU usage when compared to sequential model. Furthermore, we incorporate cellular heterogeneity and phenotypic variability by sampling the QS system parameters from Gaussian distribution. It is noteworthy that existing literature on QS [2,14] has modeled up to a population of 240 cells, whereas our proposed framework has been used to simulate a population of 2000 cells. Scalability experiments are performed on 50 cores of Forge high performance computing clusters built on Rocks 6.1.1, while other experiments are performed in Ubuntu 14.04 system with 8 CPUs.

This paper is organized as follows. Section 2 presents an overview of the overall QS system. Section 3 discusses the details of the Gillespie algorithm, sequential model and the parallel QS framework. Section 4 compares the experimental results. Finally, Sect. 5 closes the paper with concluding remarks.

2   System Overview

Our Quorum Sensing (QS) system consists of a population of cells and their shared environment, defined as the concentration of autoinducers outside the cells (external autoinducer). Figure 1 shows a population of cells and LuxI/LuxR regulatory network within bacterium *Vibrio fischeri* [3].

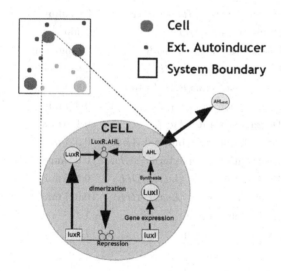

Fig. 1. Population of cells and macro view of each cell with the LuxI/LuxR regulatory network

In a QS system, bacteria communicate with each other through autoinducers (AHL), produced due to the synthesis of protein LuxI. These molecules are small in size and can diffuse freely through cell membrane into the environment and from the environment back into cell. The diffusion process changes the concentration of environment, which in turn, affects the whole population. The overall system remains coupled through this diffusion process. Within a cell, the LuxR protein binds with AHL to form the monomer (LuxR.AHL). Then, through dimerization, this monomer forms (LuxR.AHL)$_2$, repressing transcription of luxI gene. Diffusion rate of intra-cellular autoinducer AHL depends on the concentration of AHL itself, and extra-cellular autoinducer (AHL$_{ext}$). The chemical reactions and corresponding rate constants are taken from [3].

Below, we have provided a list of the chemical reactions (in accordance with the reduced QS model discussed in [3]). The propensity of the LuxI expression reaction is shown in Eq. 1 and the rate constant parameters for the reactions are listed in Table 1.

Table 1. List of constant parameters [3] used in our framework

Parameter	Description	Value
tt_{LuxR}	Protein expression rate: $LuxR$	76 copies/min
tt_{LuxI}	Protein expression rate: $LuxI$	219 copies/min
k_{-1}	Dissociation rate: $LuxR$ to AHL	$10\,\text{min}^{-1}$
k_{-2}	Dissociation rate: $(LuxR.AHL)_2$	$1\,\text{min}^{-1}$
α	Basal expression rate: $luxI$	0.01
k_A	Synthesis rate: AHL by $LuxI$	$0.04\,\text{min}^{-1}$
D	Diffusion rate: AHL	$2\,\text{min}^{-1}$
k_{d1}	Dissociation const.: $LuxR$ to AHL	100 molecule
k_{d2}	Diss. const.: $(LuxRAHL)$	20 molecule
k_{dlux}	Diss. const.: $(LuxRAHL)$ to lux	100 molecule
d_I	Degradation rate: $LuxI$	$0.027\,\text{min}^{-1}$
d_R	Degradation rate: $LuxR$	$0.156\,\text{min}^{-1}$
d_A	Degradation rate: internal AHL	$0.057\,\text{min}^{-1}$
d_{A_e}	Degradation rate: external AHL	$0.04\,\text{min}^{-1}$
d_{RA}	Degradation rate: $(LuxR.AHL)$	$0.156\,\text{min}^{-1}$
d_{RA_2}	Degradation rate: $(LuxR.AHL)_2$	$0.017\,\text{min}^{-1}$
V_{cell}	Initial cell volume	$1.1e^{-9}\,\text{l}$
V_{ext}	Extracellular volume	$1.1e^{-3}\,\text{l}$

$$\text{LuxI} \xrightarrow{\text{d}_\text{I}} \phi$$

$$(\text{LuxR} \cdot \text{AHL})_2 \xrightarrow{\text{f}(\text{x}_3,\text{t})} \text{LuxI} + (\text{LuxR} \cdot \text{AHL})_2$$

$$\text{LuxR} \xrightarrow{\text{d}_\text{R}} \phi$$

$$\xrightarrow{\text{tt}_\text{LuxR}} \text{LuxR}$$

$$\text{LuxR} + \text{AHL} \xleftrightarrow[\text{K}_{-1}]{\text{k}_{-1}/\text{k}_{\text{d1}}} \text{LuxR} \cdot \text{AHL}$$

$$(\text{LuxR} \cdot \text{AHL})_2 \xrightarrow{\text{d}_{RA}} \phi$$

$$2\,(\text{LuxR} \cdot \text{AHL}) \xleftrightarrow[K_{-2}]{k_{-2}/k_{d2}} (\text{LuxR} \cdot \text{AHL})_2$$

$$\text{AHL} \xrightarrow{\text{d}_A} \phi$$

$$\text{LuxI} \xrightarrow{k_A} \text{LuxI} + \text{AHL}$$

$$\text{AHL} \xleftrightarrow[DV_C]{D} \text{AHL}_{\text{ext}}$$

$$\text{AHL}_{\text{ext}} \xrightarrow{\text{d}_{Ae}} \phi$$

$$f(x_3, t) \triangleq tt_{LuxI}\left(\frac{k_{dlux} + \alpha_i x_3}{k_{dlux} + \alpha_i}\right) \tag{1}$$

3   Sequential and Parallel QS Frameworks

In this section we discuss the details of the sequential model for QS and the proposed parallel framework.

3.1   System Variables

Let us consider a system with χ reactions. We define the reactions set $\Gamma = \{\gamma_i \,|\, i \in \mathbb{N}, i \le \chi\}$, set of reaction rate constants $K = \{k_i | i \in \mathbb{N}, i \le \chi\}$, and the concentration of jth reactant in reaction γ_i as ω_i^j. We consider a molecular concentration matrix $M_{n \times m}$, where n and m are the number of cells and molecule species in the QS system, respectively. Thus, M_i denotes molecular concentration vector of the ith cell. Thus, $M_{i,5}$ stores the AHL_{ext} concentration for i^{th} cell. Since AHL_{ext} is a global system variable, $M_{i,5}$ remains same for all cells.

3.2   Gillespie Algorithm

The Gillespie Algorithm, also called Stochastic Simulation Algorithm (SSA), is a procedure for simulating changes in the molecular concentration of chemical species in a chemically reacting system. Hence the behavior of each cell and the advancement of simulation time within the QS system is determined by executing the Gillespie algorithm. The Gillespie algorithm (shown below) calculates *propensity* (likelihood) a_i of reaction γ_i, and the *total propensity* A as follows:

$$a_i = k_i \times \prod_{j=1}^{|\omega_i|} \omega_i^j \tag{2}$$

$$A = \sum_{i \in \Gamma} a_i \tag{3}$$

In each step, the Gillespie algorithm takes four input parameters: the set of reaction rules (R_r), reaction rate constant (R_c), initial simulation time t

and molecular concentration matrix M. The Gillespie algorithm probabilistically chooses a single reaction (γ_j), based on the individual reaction propensities. Finally, Gillespie updates the reactant concentration of γ_j according to its stoichiometric coefficients.

```
1: procedure GILLESPIE(Rr,Rc,M,t)
2:     Calculate ai for all reactions and A.
3:     Select r1,r2 = random(0,1)
4:     Update current time t: t = t + ln(1/r1)/A
5:     Select reaction γj :    (∃J ∈ ℕ) Σ^J_{j=1} aj < A × r2 ≤ Σ^{J+1}_{j=1} aj
6:     Update reactant concentration of γj in M.
7:     Return t, M
8: end procedure
```

3.3 Sequential Model

In the sequential QS model, each cell contains one copy of the regulatory network (as shown in Fig. 1). All the reactions in the entire population, including diffusion reaction between cell and environment, is considered as a single global system. As shown in Line 4 of sequential algorithm, the Gillespie algorithm is invoked in a loop, until the current time t exceeds the total simulation time T. Note that the sequential QS, when used in conjunction with the exact Gillespie algorithm, yields accurate result. Therefore, we implement the sequential model as a benchmark of accuracy for our proposed parallel QS framework.

```
1: procedure SEQUENTIAL(Rr,Rc,M,T)
2:     t = 0
3:     While t ≤ T do
4:         t, M = Gillespie(Rr,Rc,M,t)
5:     Endwhile
6: end procedure
```

3.4 Parallel Framework

The parallel QS framework is implemented using the Multiprocessing library of Python [15]. Here we discuss the different aspect of the parallel framework.

Steps in Parallel Framework: As shown in Fig. 2, the virtual master process takes 3 inputs: reaction rules (R_r), environment and constant (R_c). (1) The master spawns several parallel processes. Here, each cell is modeled as a memoryless process, termed *cell process*. In a large scale system several cell processes are assigned to a single core. (2) In each time intervals, master sends molecular concentration and environment information to each cell process and invokes Gillespie Algorithm. (3) Each cell process runs the Gillespie algorithm locally

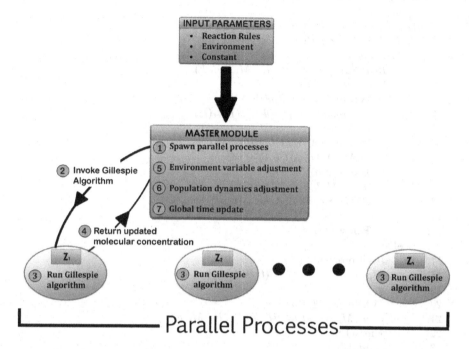

Fig. 2. Overview of steps in the parallel QS framework

and (4) returns the updated molecular concentration to the master. Following this, the master node (5) updates environment variable, (6) population dynamics and (7) global time, before returning to first step. This cycle continues until simulation duration is reached.

Master Process: The master process reads and sends sampling interval Ψ, R_r, R_c and M_i to all the cell processes, at each time instance t. *Sampling interval (Ψ) is defined as the time interval between which the master process collects concentration data from all the cell processes.* After each Ψ interval, the master receives the updated M_i from each Z_i. It is noteworthy that in the sequential approach, a single reaction takes place at a time. Consequently, any change to AHL_{ext} is instantly reflected to $M_{i,5}$ (for all i). However, in the parallel approach, each cell process Z_i autonomously invokes Gillespie and updates $M_{i,5}$. Thus, we ensure uniformity of AHL_{ext} in the system using Eq. 4.

Cell Processes: Each cell process Z_i invokes the Gillespie algorithm and returns updated M_i to the master. Let us discuss two key aspects of the ParallelQS algorithm.

```
 1: procedure PARALLELQS()
 2:    If ID = Master then
 3:        t = 0
 4:        ReadParameters(R_r, R_c, M, T, n, Ψ)
 5:        While t ≤ T do
 6:            While (not sentToAll()):
 7:                send(M_i, R_r, R_c, Ψ, AHL_ext(t))
 8:            Endwhile
 9:            While (not recvFromAll()) do
10:                M_i, AHL^i_ext(t + 1) = recv()
11:            Endwhile
12:            J = {AHL^i_ext(t + 1)|i ≤ n}
13:            ADJUSTEXT(AHL_ext(t), J)
14:            t = t + Ψ
15:        Endwhile
16:    else
17:        (Cell process)
18:        M_i, R_r, R_c, Ψ, AHL_ext(t) = recv()
19:        gt = 0
20:        While gt ≤ Ψ do
21:            gt, M_i = GILLESPIE(R_r, R_c, M_i, gt)
22:        Endwhile
23:        send(M_i)
24:    Endif
25: end procedure
```

Time Synchronization: In each time instance, the master must wait till it has heard from all cell processes (Z_is). Following this, it increments the overall system time t by sampling interval $Ψ$.

Time Increment: Gillespie calculates the increment in simulation time t as $ln(1/r_1)/A$, where r_1 is random number between 0 and 1. Thus, there exists an inverse relationship between time variable t and overall propensity A (i.e., $t \propto \frac{1}{A}$). Note that the sequential model calculates A (Eqs. 2 and 3) based on all reactions in the system, whereas each Z_i in the parallel framework calculates A_i only based on the reactions specific to itself, implying that $A_i \ll A$. Hence, the time increment for the parallel framework is expected to be greater than that of the sequential approach.

Adjustment of Global AHL_{ext} Concentration: We assume that AHL_{ext} is homogeneously distributed within the system boundary, making it a global system variable. A cell is capable of interacting with any AHL_{ext} molecule. Master estimates the adjusted system AHL concentration at time $t + 1$, $AHL_{ext}(t + 1)$, by incrementing $AHL_{ext}(t)$ by the net difference of $AHL_{ext}(t)$ concentration from local cellular AHL concentrations, over all cells, using the equation below:

$$AHL_{ext}(t+1) = AHL_{ext}(t) + \sum_{j=2}^{n+1}(AHL_{ext}^j(t+1) - AHL_{ext}(t)) \qquad (4)$$

Here $AHL_{ext}^j(t+1)$ is the local AHL concentration of the jth cell process at time $t+1$.

Adjustment of Molecular Concentration Due to Population Dynamics. The proposed framework is capable of modeling dynamic population (i.e. cellular birth and death). Each cell Z_i undergoes division once simulation time t exceeds its division time β_i. For each Z_i, we sample β_i from exponential distribution with $\mu = 45$ min. Each cell division causes the mother cell to split into two daughter cells, each containing half the molecular concentration of mother cell. Given initial volume of all cells V_{cell}, each Z_i has a *time-dependent volume*, given by –

$$V_i(t) = V_{cell} \times 2^{\frac{t}{\beta_i}} \qquad (5)$$

Initial volume of a daughter cell is half the volume of its mother cell. The propensity of each reaction is also updated to account for time-dependent volume. In our QS system, cell density is kept constant by compensating each cell division by death of randomly picked cell (as discussed in [2]).

4 Results

We consider 6 molecule species in our reduced QS system – **(A)** LuxI, **(B)** LuxR, **(C)** (LuxR.AHL)$_2$, **(D)** AHL **(E)** LuxR.AHL and **(F)** AHL$_{ext}$. Parallel framework is implemented using Python 2.7 and Python Multiprocessing library [15]. Scalability experiments are performed on 50 cores of Forge high performance cluster; for other simulation experiments, we use Ubuntu system with Linux system with 8 CPUs of 1.6 GHz each.

4.1 Accuracy

We analyze how closely the dynamics of molecular concentration (matrix M) generated by parallel QS framework aligns with that of the sequential model. On both *sequential and parallel approaches*, we simulate 50 cells for 200 min and sampling interval $\psi = 0.05$.

Similarity of Molecular Concentration. Fig. 3 shows that the average concentration dynamics of molecules (A)–(F) (over 10 trials) are nearly identical for sequential and parallel approaches.

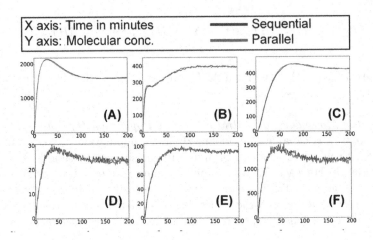

Fig. 3. Comparison of average molecule concentration dynamics of 6 molecular species – (A) LuxI, (B) LuxR, (C) $(\text{LuxR.AHL})_2$, (D) AHL (E) (LuxR.AHL) and (F) (AHL_{ext}) for population of 50 cells.

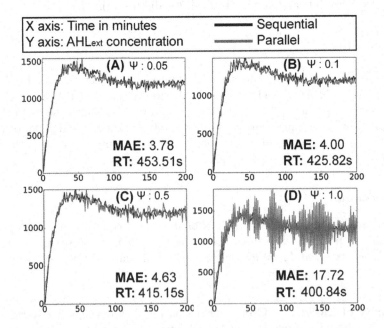

Fig. 4. Increase in sampling interval (Ψ) increases Mean Absolute Error (MAE), but decreases running time (RT), between sequential and parallel QS framework.

Effect of Sampling Interval. The choice of Ψ in the parallel framework affects the number of the data points master receives from each $Z[i]$. If Ψ is high, we expect our framework to exhibit inaccuracy in concentration dynamics due to lack of sufficient data. Here we consider *Mean Absolute Error (MAE)* as our metric of accuracy i.e. higher the MAE between two sets of plot points, greater is the dissimilarity. Since the sequential model is our benchmark, we calculate the deviation of AHL_{ext} of the parallel framework (ϕ_p) from the sequential model (ϕ_S) on d data points, as $MAE(\phi_s, \phi_p) = \frac{1}{d} \times \sum_d |M_{i,5}^s(d) - M_{i,5}^p(d)|$.

Simulation on 50 cells for duration of 200 min shows that the MAE of AHL$_{ext}$ plots between the two approaches increases with the increase in Ψ (Fig. 4). We observe that a higher Ψ, though enhances running time (RT) by minimizing communication overhead, degrades accuracy.

4.2   Population Dynamics

We model cellular birth and death in QS for a population size 1000 cells and duration 100 min. In Fig. 5(A), a single color represents the LuxI concentration of a bacterial cell over time, while the discontinuity and drop in the curves show cell death and division, respectively.

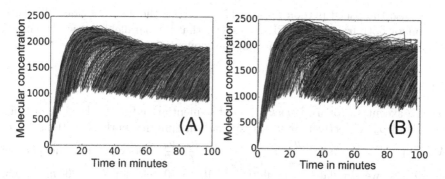

Fig. 5. Population dynamics: LuxI concentration during cellular birth and death under conditions of (A) no noise and (B) noise with standard deviation 0.005.

Our proposed parallel framework is capable of modeling noise arising from stochastic fluctuation in gene expression that can cause phenotypic variability in isogenic population [16]. In Fig. 5(B), we show cellular birth and death under condition of noise generated by sampling constant parameters R_C from Gaussian distribution with standard deviation 0.005. It is noteworthy that molecular concentration in Fig. 5(B), by virtue of the noise, exhibits greater phenotypic variability than the one in Fig. 5(A).

4.3 Processor Utilization

We compare the CPU utilization of the sequential and parallel frameworks on a population size of 100 cells and duration of 60 min on an 8-CPU machine.

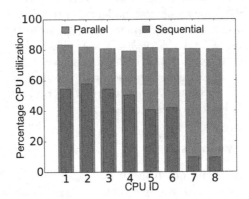

Fig. 6. Comparison of CPU utilization for the sequential and parallel QS frameworks

We use the psutil python library [17] to record the instantaneous CPU utilization for both frameworks. Figure 6 shows that the parallel framework exhibits a more uniform CPU utilization than the sequential approach.

4.4 Speed up

Let the execution (or *wall clock*) time of sequential and parallel QS algorithms be ρ_s and ρ_p respectively, and the simulation time for both algorithms be T. We define speedup $S_p = \dfrac{\rho_s}{\rho_p} = \dfrac{\rho_s}{T} \times \dfrac{T}{\rho_p} = \dfrac{\rho_s}{T} / \dfrac{\rho_p}{T}$. For sequential and parallel algorithms, we simulate a population of 10 to 50 cells for $T = 200$ min each. Given that T is same for both algorithms, we compare their real (or execution) time to analyze the speedup rendered by parallel QS framework. We generate semi-log plots for execution time for population sizes 10, 20, 30, 40 and 50. Figure 7(A) shows that our framework incurs extremely little increase in real time with growth in cell population size as compared to the sequential approach.

4.5 Scalability

Finally, we compare the sequential and parallel approaches on the basis of execution time for population sizes 25, 50, 100, 500, 1000 and 2000 cells. Figure 7 and (B) is a log-log plot showing that the parallel framework incurs significantly lower execution time. It is noteworthy that the sequential approach does not scale beyond 100 cells, thus the expected real time values for sequential approach (shown in dotted red line) is obtained through extrapolation.

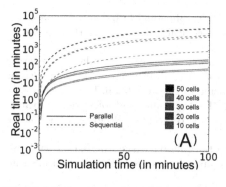
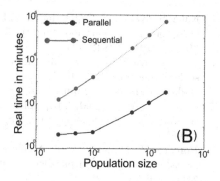

Fig. 7. (A) Speed-up: Semi-log plot for speed-up for population sizes varying between 10 and 50. Parallel QS framework exhibits little growth in running time with increase in simulation time (B) Scalability: Log-log plot for scalability analysis for population sizes 25, 50, 100, 500, 1000, 2000. (Color figure online)

5   Conclusion and Future Work

In this paper we present a scalable parallel framework for QS in bacteria. Simulation for varying population size, sampling interval and duration show that our framework models concentration dynamics almost as accurately as its sequential counterpart, while showing significant improvement in speed-up and CPU utilization. We also study the degradation in accuracy and improvement in running time of this model, with increase in sampling interval. We show that the framework incorporates cellular heterogeneity, phenotypic variability and scalability by sampling the QS system parameters from Gaussian distribution and modeling up to 2000 bacterial cells. Our future works involve the extension of the framework to accommodate spatial positioning of cells and environmental molecules to incorporate cellular heterogeneity. In the spatial model, the master process will recalculate the coordinates of the cells and environmental molecules after each sampling interval.

Acknowledgement. The research presented in this work is supported by the National Science Foundation CBET-CDS&E grant 1609642.

References

1. Miller, M.B., Bassler, B.L.: Quorum sensing in bacteria. Ann. Rev. Microbiol. **55**(1), 165–199 (2001)
2. Weber, M., Buceta, J.: Dynamics of the quorum sensing switch: stochastic and non-stationary effects. BMC Syst. Biol. **7**, 6 (2013)
3. Boada, Y., Vignoni, A., Pico, J.: Promoter and transcription factor dynamics tune protein mean and noise strength in a quorum sensing-based feedback synthetic circuit. bioRxiv, p. 106229 (2017)
4. Bassler, B.: Manipulating quorum sensing to control bacterial pathogenicity. FASEB J. **29**(1 Suppl.), 88-1 (2015)

5. Kang, S., Kahan, S., McDermott, J., Flann, N., Shmulevich, I.: Biocellion: accelerating computer simulation of multicellular biological system models. Bioinformatics **30**(21), 3101–3108 (2014)
6. Tian, T., Burrage, K.: Parallel implementation of stochastic simulation for large-scale cellular processes. In: Proceedings of the Eighth International Conference on High-Performance Computing in Asia-Pacific Region, pp. 6–pp. IEEE (2005)
7. Li, H., Cao, Y., Petzold, L.R., Gillespie, D.T.: Algorithms and software for stochastic simulation of biochemical reacting systems. Biotechnol. Progress **24**(1), 56–61 (2008)
8. Komarov, I., D'Souza, R.M.: Accelerating the gillespie exact stochastic simulation algorithm using hybrid parallel execution on graphics processing units. PloS One **7**(11), e46693 (2012)
9. Harvey, D.G., Fletcher, A.G., Osborne, J.M., Pitt-Francis, J.: A parallel implementation of an off-lattice individual-based model of multicellular populations. Comput. Phys. Commun. **192**, 130–137 (2015)
10. Kouskoumvekakis, E., Soudris, D., Manolakos, E.S.: Many-core CPUs can deliver scalable performance to stochastic simulations of large-scale biochemical reaction networks. In: 2015 International Conference on High Performance Computing & Simulation (HPCS). IEEE (2015)
11. Vanneschi, M.: The programming model of assist, an environment for parallel and distributed portable applications. Parallel Comput. **28**(12), 1709–1732 (2002)
12. Dematté, L., Mazza, T.: On parallel stochastic simulation of diffusive systems. In: Heiner, M., Uhrmacher, A.M. (eds.) CMSB 2008. LNCS (LNAI), vol. 5307, pp. 191–210. Springer, Heidelberg (2008). https://doi.org/10.1007/978-3-540-88562-7_16
13. Islam, M.A., Roy, S., Das, S., Barua, D.: Multicellular models bridging intracellular signaling and gene transcription to population dynamics. Processes **6**(11), 217 (2018)
14. Boada, Y., Vignoni, A., Navarro, J.L., Picó, J.: Improvement of a CLE stochastic simulation of gene synthetic network with quorum sensing and feedback in a cell population. In: 2015 European Control Conference (ECC), pp. 2274–2279. IEEE (2015)
15. Oudkerk, R., Noller, J.: Multiprocessing process-based threading interface. https://docs.python.org/2/library/multiprocessing.html
16. Boada, Y., Vignoni, A., Picó, J.: Engineered control of genetic variability reveals interplay among quorum sensing, feedback regulation, and biochemical noise. ACS Synth. Biol. **6**(10), 1903–1912 (2017)
17. Rodola, G.: Psutil: cross-platform lib for process and system monitoring in python. https://pypi.org/project/psutil/

Membrane Computing Aggregation (MCA): An Upgraded Framework for Transition P-Systems

Alberto Arteta[1(✉)], Luis Fernando Mingo[2], Nuria Gomez[2], and Yanjun Zhao[1]

[1] Computer Science, Troy University, Troy, USA
{aarteta,yjzhao}@troy.edu
[2] Computer Science, Polytechnic University of Madrid,
Crtra Valencia km 7, Madrid, Spain
{lfmingo,ngomez}@eui.upm.es

Abstract. MCA (Membrane computing aggregation is experimental computational frame. It is inspired by the inner properties of membrane cells (Bio-inspired system). It is capable of problem solving activities by maintaining a special, "meaningful" relationship with the internal/external environment, integrating its self-reproduction processes within the information flow of incoming and outgoing signals. Because these problem solving capabilities, MCA admits a crucial evolutionary tuning by mutations and recombination of theoretical genetic "bridges" in a so called "aggregation" process ruled by a hierarchical factor that enclosed those capabilities. Throughout the epigenetic capabilities and the cytoskeleton and cell adhesion functionalities, MCA model gain a complex population dynamics specifics and high scalability. Along its developmental process, it can differentiate into meaningful computational tissues and organs that respond to the conditions of the environment and therefore "solve" the morphogenetic/configurational problem. MCA, above all, represents the potential for a new computational paradigm inspired in the higher level processes of membrane cells, endowed with quasi universal processing capabilities beyond the possibilities of cellular automata of and agent processing models.

Keywords: Membrane system · Membrane computing · Natural computing

1 Introduction

In spite of all the recent emphasis and advancements in systems biology, synthetic biology, and network science about modelling of gene networks, protein networks, metabolic and signaling networks, etc. some of the most important computational properties of membrane cells have not been grappled and "abstracted" et: scalability, tissular differentiation, and morphogenesis - i.e., the capability to informationally transcend the cellular level and organize higher level information processes by means of heterogeneous populations of membrane cells organized as "computational tissues and organs".

A. Compagnoni et al. (Eds.): BICT 2019, LNICST 289, pp. 195–207, 2019.
https://doi.org/10.1007/978-3-030-24202-2_15

Synthetic biology has become extraordinarily active in the manufacture of very simple and robust models and simulations tailored to the realization problems of circuits and modules in vivo, mostly addressed to prokaryotic systems. In the first wave of these studies, very basic elements such as promoters, transcription factors, and repressors were combined to form small modules with specified behaviors. Currently modules include switches, cascades, pulse generators, oscillators, spatial patterns, and logic formulas [1]. The second wave of synthetic biology is integrating basic parts and modules to create systems-level circuitry. genomes and synthetic life organisms are envisioned, and application-oriented systems are contemplated. Different computational tools and programming abstractions are actively developed (the Registry of Standard Biological Parts; the Growing Point Language GLP; the Origami Shape Language OSL, the PROTO bio programming language, etc. See details at the Open Wetware site). Evolving cell models of prokaryotes have also been addressed [2, 3]. As some have put, "systems broaden the scope of synthetic biology designing synthetic circuits to operate in reliably in the context of differentiating and morphologically complex membrane cells present unique challenges and opportunities for progress in the field" [4]. However, very few synthetic biology researchers do contemplate using systems.

In systems biology, a plethora of modelling developments have been built around signaling pathways, cell cycle control, topologies of protein networks, transcriptional networks, etc. There is a relatively well consolidated thinking, in part due to traditional physiology and to systems science and control theory which were at the origins of this new field, of going "from genes to membrane cells to the whole organ" as D. Noble has done for heart models [5]. The integration of proteins to organs has also been promoted by bioinformatic-related projects such as the "Physiome Project" [6]. Important works have been done in the vicinity of "network science" in order to make sense of gene networks, protein networks, transcription networks, complexes formation, etc. For instance, about how is dynamically organized modularity in the yeast protein-protein interaction network [7], it was uncovered that two types of "hub" contribute to the organized modularity of the proteome: "party" hubs which interact with their partners simultaneously, and "date" hubs, which bind their different partners at different times and locations (we will see later on the importance of the discussion on "modularity" in the evo-devo field). Predictive models of mammalian membrane cells have been described using graph theory, assembling networks and integrative procedures [8]. Important systems biology compilations and far-reaching cellular models have been made by [9], Kitano [10–12] It has to be emphasized that concerning the views advocated in this proposal, most of systems biology works depart from the goal of "abstracting computational power out from systems" and focus instead on "applying computational power to analyze the organization of systems." Notwithstanding the foregoing, studies such as A. [13] on bacteria as computers making computers [14], on the operating system of bacteria could be considered as forerunners in the former direction.

In the science of development (the "evo-devo" discipline) most of the emphasis has been on modularity. What it exactly means in developmental terms is still a matter of controversy [15–17]; but undoubtedly modularity refers to the capability of cellular networks to dissociate networked processes at a lower level and to recombine or redeploy them at the higher level of the multicellular organism. Thanks to the cellular signaling system, the genetic switches, the cytoskeleton, and some other topobiological mechanisms [18], the unitary network of cellular processes integrated into the cell-cycle may be broken down into coherent modules and be performed separately in different membrane cells within differently specialized tissues [19]. This implies a flexible organization for the deployment of biomolecular processing modules, which actually are "cut" differently in each tissue along the developmental process, due also to chromatin remodelling during development [20]. Interestingly, not only differentiation but also morphology becomes an instance of the scalable "modular" processing, throughout the "tensegrity" emergent property and the ontogenetic arrangement of symmetry breakings in a force field. The emergence of cellular bauplans where sig-naling, force fields, and cytoskeletal mechanical modes conspire together to create but a few basic morphologies for membrane cells, depending also on the populations present, seems to be another important consequence [21]. Interestingly, complex morphologies obtained out from Turing diffusion model have been cogently discussed as a result of cell-to-cell developmental interactions [22]. Currently, the evo-devo field accumulates a considerable mass of biomolecular-organization-facts, poorly conceptualized yet, to be computationally "abstracted" in the perspective of MCA advancement.

In the fields closer to computer science and Biocomputing, it has been important the introduction of the agent based approach (as pioneered by W. Fontana and others), which uses sets of rules to define relationships between cellular components substi-tuting for the simple Boolean networks and differential equations used up to now. Proteins and other biomolecules become molecular "automata" and the aggregate behavior that emerges out from these models is the combinatorial expression of all those automata doing their specific micro-functions [23]. This approach shows promise for "evolvable" advancement of network models endowed with the flexible modularity property. It is somehow close to the already mentioned predictive models of mam-malian membrane cells that are using graph theory, assembling networks and inte-grative procedures [24]. New generations of cellular models (of "automata") have been developed too, with powerful data content and with potential for modelling multi-cellular systems in a general way, supporting user-friendly in silicon experimentation and discovery of emergent properties [25]. Under the approach of Artificial Embry-ology, a developmental system has been obtained by means of cellular automata systems capable of following "rewriting rules" procedures, emulating elementary morphologies and multicellular distributions [26].

As for the developments in molecular Biocomputing, the idea that bio-molecules (DNA, RNA, proteins) might be used for computing already emerged in the fifties and was reconsidered periodically with more and more arguments which made it more viable. But the definitive confirmation came in 1994 [27] when in [27] successfully accomplished the first experimental close connection between molecular biology and computer science. He described how a small instance of a computationally intractable problem might be solved via a massively parallel random search using molecular biology methods. An important part of this project is focusing on bio-inspired models of computation abstracted from the very complex networks in living systems. Its goal is to investigate several aspects of these models particularly focused on connections between theoretical models and natural (biological) networks.

The main topics are: Computational aspects (computational power, structural and description complexity).

Several new directions of research have been initiated in the last decade: computing devices inspired from the genome evolution [28–30] with an explosive development, evolutionary systems based on the behavior of cell populations [31]) computing models simulating the process of gene assembly in ciliates Swarm computation is mainly based on the same idea: a swarm is a group of mobile biological organisms wherein each individual communicates with others by acting on its local environment [32]. Regarding applicative models there are many attempts to update Cells computing paradigm in [33–36] among others.

2   Membrane Computing

A Transition P System of degree n $n > 1$ is a construct $\prod = (V, \mu, \omega_1, .., \omega_n, (R_1, \rho_1), ..(R_n, \rho_n), i_0)$

Where:

V is an alphabet; its elements are called objects;

μ is a membrane structure of degree n, with the membranes and the regions labeled in a one-to-one manner with elements in a given set; in this section we always use the labels 1, 2, n;

$\omega_i \, 1 \leq i \leq n$, are strings from V^* representing multisets over V associated with the regions $1, 2, \ldots,$ n of μ

$R_i \, 1 \leq i \leq n$, are finite set of evolution rules over V associated with the regions $1, 2, \ldots,$ n of μ; ρ_i is a partial order over $R_i \, 1 \leq i \leq n$, specifying a priority relation among rules of R_i. An evolution rule is a pair (u, v) which we will usually write in the form $u \rightarrow v$ where u is a string over V and $v = v'$ or $v = v' \, \delta$ where v' is a string over $(V \times \{here, out\}) \cup (V \times \{in_j \, 1 \leq j \leq n\})$, and δ is a special symbol not in. The length of u is called the radius of the rule $u \rightarrow v$

i_o is a number between 1 and n which specifies the output membrane of $\prod$.

Let U be a finite and not empty set of objects and N the set of natural numbers. A multiset of objects is defined as a mapping:

$$M : U \rightarrow N$$

$$a_i \rightarrow u_1$$

Where a_i is an object and u_i its multiplicity.

As it is well known, there are several representations for multisets of objects (Fig. 1).

$$M = \{(a_1, u_1), (a_2, u_2), (a_3, u_3) \ldots\} = a_1^{u_1} \cdot a_2^{u_2} \cdot a_n^{u_n} \ldots \ldots$$

Note: Initial Multiset is the multiset existing within a given region in where no application of evolution rules has occurred yet.

Definition Evolution rule with objects in U and targets in T.

Evolution rule with objects in U and targets in T is defined by $r = (m, c, \delta)$ where $m \in M(U), c \in M(U \times T)$ and $\delta \in \{to \, dissolve, \, not \, to \, dissolve\}$

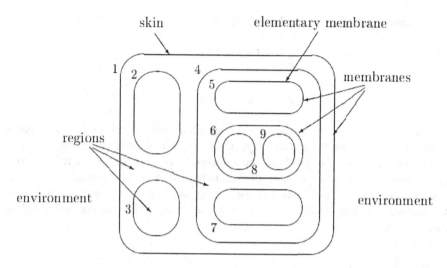

Fig. 1. P-system structure definition multiset of objects

From now on 'c' will be referred a s the consequent of the evolution rule 'r'.

Note The set of evolution rules with objects in U and targets in T is represented by R (U, T).

Definition Multiplicity of an object in a multiset of objects M(U)

Let $a_i \in U$ be an object and let $m \in M(U)$ be a multiset of objects. The multiplicity of an object is defined over a multiset of objects such as:

$$\|_{a_i}: U \times M(U) \to N$$
$$(a_i, m) \to |m|_{a_i} = n | (a_i, n) \in m$$

Definition Multiplicity of an object in an evolution rule r

Let $a_i \in U$ be an object and let $R(U, T)$ be a multiset of evolution rules. Let $r = (m, c, \delta) \in R(U, T)$ where $m \in M(U), c \in M(U \times T)$ and $\delta \in$ {*to dissolve, not to dissolve*}

The multiplicity of an object is defined over an evolution rules such as:

$$\|_{a_i}: U \times R(U, T) \to N$$
$$(a_i, r) \to |m|_{a_i} = n | (a_i, n) \in m$$

P-system evolution

Let C_i be the consequent of the evolution rule r_i. Thus, the representation of the evolution rules is:

$$
\begin{aligned}
r_1 &: a_1^{u_{11}} a_2^{u_{12}} \ldots a_n^{u_{1n}} \rightarrow C_1 \\
r_2 &: a_1^{u_{21}} a_2^{u_{22}} \ldots a_n^{u_{2n}} \rightarrow C_2 \\
&\ldots\ldots\ldots\ldots\ldots\ldots \rightarrow \ldots\ldots\ldots\ldots\ldots\ldots\ldots\ldots\ldots\ldots\ldots \\
r_m &: a_1^{u_{m1}} a_2^{u_{m2}} \ldots a_n^{u_{mn}} \rightarrow C_m
\end{aligned}
\tag{1}
$$

P-systems evolve, which makes it change upon time; therefore, it is a dynamic system. Every time that there is a change on the p-system we will say that the p-system is in a new transition. The step from one transition to another one will be referred to as an evolutionary step, and the set of all evolutionary steps will be named computation. Processes within the p-system will be acting in a massively parallel and non-deterministic manner. (Similar to the way the living cells process and combine information).

3   The Upgrade

The proposal is a new computational paradigm based on Membrane cells, scalable ones which are capable to produce "computational tissues and organs". The organization of such computational tissues and organs is inspired by the emerging informational properties of biomolecular networks and will be based on scalable "membrane cells" guided by functional rules similar to the biological ones (molecular recognition, self-assembly and topo biology-theory rules).

The direct inspiration from the membrane cells is precisely the breakthrough of the MCA project. By building computational tissues our proposal makes an evolutionary jump with respect of today research in this field, mainly focused on aggregates of unicellular organisms (e.g. bacteria). Far from modelling and simulating the cellular processes, our computational paradigm will be a clear abstraction of the basic mechanisms and computational capabilities of the membrane cells and tissues, in order to solve complex problems in a new (bioinspired) way.

Real tissues display far more complex properties (emergent properties) than the sum of the properties of the individual membrane cells they are made from. In the same way, the emergent properties and functions of our membrane cells and computational tissues will be used for the resolution of real problems, impossible to be appropriately solved by conventional methods: not only biological morphogenesis, but also evolution of economic systems and prediction of crisis, optimization of "industrial ecologies", analysis of the dynamics of social interactions and conflicts, ecosystem disturbances, etc., that are more complex than combinatorial optimization, as well as other classical NP-Complete ones.

Our "membrane cells" will be a species of "proto-membrane cells" and a far objective of the project is also the ex-novo synthesis of "membrane cells" and tissues performing as living computational biomolecular networks. The long-term vision that motivates this breakthrough is to build new information processing devices with evolving capabilities, which will adapt themselves to the complexity of the problems. In particular, we foresee a synthetic approach to build computational membrane cells and tissues, and to create computational bio-inspired devices of higher complexity (tissues-organs). A far future objective of the project goes beyond the mathematical, software and hardware tools. It is to obtain in lab synthesized "living" information processing systems based on artificial "membrane cells" and hybrid systems combining living components (our "synthesized membrane cells") and non-living elements (e.g. silicon-based).

MCA approach is the most appropriate to deal with extremely complex problems that will be crucial in the future. It shows potential to go beyond classical Biocomputing strategies such as self-reproducing machines, cellular automata, perceptron's & neural networks, genetic algorithms, adaptive computing, bacteria-based computation, artificial membrane cells, etc. Specifically, a new generation of natural computing could be built, based upon the scalable "membrane cells" with problem solving capacity in very different realms: biomaterials and bioengineering, non-linear parallel processing, design of bioinspired systems, modelling of economic, industrial and financial systems, optimization strategies in social settings, etc.

For the achievement of our long-term objectives we need to:

analyze the wide amount of existing knowledge regarding one of the deepest sources of biocomputational power, the topological and flexible networking properties of biomolecular scalable modules in membrane cells,

realize an abstraction of the basic mechanisms and computational capabilities of the membrane cells both at sub cellular and networking level, and develop formal models to be used in new information processing technologies, basically based on combinatory processes of protein domains and genetic switches, together with cytoskeleton dynamics and topobiology-theory,

use the above proposed models to create scalable "/proto membrane cells" and abstract-formal "evolvable" cellular networks and computational tissues & organs endowed with these flexible modularity properties.

For our far final objective we need to obtain in lab proof that synthesis of new forms of living "membrane cells" in an inverse process: "membrane cells and tissues" => "theoretical abstract/formal models" => "artificial membrane cells and tissues" => "in lab synthesized living membrane cells" is possible. MCA breakthrough is an essential step towards the achievement of our long-term vision because it will set the theoretical basis and develop the experimental tools for the creation of the scalable membrane cells, computational tissues and organs (both abstract and living ones).

4 MCA System

A MCA is a set and a set of aggregation rules among membranes. The set of aggregation rules are not fully integrated with the evolution rules of a given p-System but establishes the correlation between 2 given membrane models by deciding the way 2 or mere P-systems are being aggregated. The rules can be defined as a Matrix relation

$$\varphi_1(k_1, k_2, .., k_m,) \equiv \begin{pmatrix} u_{11} & u_{21} & ...u_{m1} \\ u_{12} & u_{22} & ...u_{m2} \\ ... & ... & ... \\ u_{1n} & u_{2n} & ...u_{mn} \end{pmatrix} \begin{pmatrix} k_1 \\ k_2 \\ ... \\ k_m \end{pmatrix} = \begin{pmatrix} u_1 \\ u_2 \\ ... \\ u_n \end{pmatrix} \tag{2}$$

Where $\varphi1(k)$ is the aggregation relation and is defined by the association of n P-systems, k determines the aggregation rules of each component in every p-system I and U are the component (objects).

Evolution rule application phase

This phase is the one that has been implemented following different techniques.

In every region within a p-system, the evolution rules application phase is described as follows:

Rules application to a multiset of object in a region is a transforming process of information which has input, output and conditions for making the transformation.

Given a region within a p-system, let $U = \{a_i | 1 \leq i \leq n\}$ be the alphabet of objects, m a multiset of objects over U and R(U, T) a multiset of evolution rules with antecedents in U and targets in T.

The input in the region is the initial multiset m.

The output is a maximal multiset m'.

The transformations have been made based on the application of the evolution rules over m until m' is obtained.

Application of evolution rules in each region of P systems involves subtracting objects from the initial multiset by using rules antecedents. Rules used are chosen in a non-deterministic manner. This phase ends when no rule is applicable anymore.

The transformation only needs rules antecedents as the consequents are part of the communication phase.

Observation

Let $k_i \in N$ be the number of times that the rule r_i is applied. Therefore, the number of symbols a_j which have been consumed after applying the evolution rules a specific number of times will be:

$$\sum_{i=1}^{m} k_i \cdot u_{ij} \tag{3}$$

Definition

Given a region R and alphabet of objects U, and R (U, T) set of evolution rules over U and targets in T.

$$\begin{aligned} r_1 &: a_1^{u_{11}} a_2^{u_{12}} ... a_n^{u_{1n}} \rightarrow C_1 \\ r_2 &: a_1^{u_{21}} a_2^{u_{22}} ... a_n^{u_{2n}} \rightarrow C_2 \\ &... \qquad\qquad \rightarrow .... \\ r_m &: a_1^{u_{m1}} a_2^{u_{m2}} ... a_n^{u_{mn}} \rightarrow C_m \end{aligned} \tag{4}$$

Maximal multiset is that one that complies with:

$$\bigcap_{l=1}^{m} \left[\bigcup_{i=1}^{n} \left(u_i - \sum_{j=1}^{m} (k_j \cdot u_{ji}) \leq u_{li} \right) \right] \tag{5}$$

5 Correction

The correction of the system fully relies in the correction of the internal P-system of the MCA. In order to prove the aggregation system is distributed then 2 processes need to be proven.

Correction of the formal definition of Transition P-System (Paun 1998)
Correction of the aggregation rules applying to 2 given P-systems.

The correction of the second point gets reduced to a deductive demonstration where the aggregation of 2 given P-systems is base case and the generic case of n-P-systems can be seen as the aggregation of n-1 P-systems (inductive case) with a correct aggregation to the last one.

Thus, the key is to prove that aggregation of 2 given P-system is a correct process and indeed reinforce the idea of full inherent parallelism and nondeterministic modelling that membrane models are after.

Aggregation rule. Let us use a short definition of a given P-System

$$\prod = \left(V, \mu, \omega_1, .., \omega_n, (R_1, \rho_1), .. (R_n, \rho_n), i_0 \right)$$

Base case. Given 2 Transition P-system

Aggregation where P_1, P_2 are 2 given P-Systems, P_{12} is the aggregated P-system where is the aggregated alphabet of both P-systems, μ_{12} is the set of regions in the aggregated P-system and ω_{12}, R_{12} are the multiset of objects and set of evolution rules of the aggregated P-system.

Building the aggregated alphabet is obvious. The result is the Union of both. Correctness for this operation is also obvious.

The aggregation of the 2set of multisets is obvious. The result is the Union of both. Correctness for this operation is also obvious.

The aggregation of the 2set of the set of the evolution rules R_{12} is obvious. The result is the Union of both. Correctness for this operation is also obvious.

There are 2 factors in the aggregation that are not obvious which are the aggregated Set of regions μ_{12}. This set of regions is constructed in our proposal as supervised and directed by the factor λ that defines the capabilities previously mentioned. This λ is defined dynamically by the nature of problem the MCA is about to fix. i.e. in a problem of sum of squares is not necessary aggregation as 2 independent P-system could calculate their squares [Paun 2001] and send those outputs to a third (obvious) one that calculates the sum of both results. However, for didactic purposes and aggregated solution could be provided in where a MCA is created with 2 Input P-systems. The aggregated would assign equal λ (priority) to both of them, and then either of them could contain the other one. The container P-system process the output of the contained P-system by adding it to an another square number.

Other problems, especially those that requires sub solutions that are part of optimization techniques would be required to establish a clear hierarchy in the aggregation of MCA. Thus:

The aggregation of the regions of 2 P-systems would be determined by a priority or hierarchy described by λ. This is a dynamic factor that must be configured right before the problem is dealt with.

The aggregated P-system will have to work the communication phase after every evolutionary step. This communication phase also fully relies on the hierarchy establish by λ and will operate as normal when the aggregation is complete and the MCA is finished.

Inductive case:

Given a successful aggregation (MCA) of n P-systems MCA(n), is it correct to aggregate n+1 P-systems?

The inductive case is a direct consequence of the aggregated property.

MCA (n) system becomes a complex P-System with an aggregation of regions according to the λ factor. MCA (n) = let's call the aggregated P-system as. Once the aggregation is seen as a P-system, aggregating it with another P_1 is obvious by applying the base case.

Simulations and results:

We have been performing some simulations in simple problem solving in same traditional computing paradigm For small problems clearly aggregation is not necessary, although the advantage of this proposal shows up, when the complexity of the problem increases. Theoretically a fully and corrected aggregated Solution (A whole MCS) would overweight the cost of the calculation of λ and he redesign of the membrane system that can always occur during compiling time anyways (Table 1).

Table 1. Comparison traditional P-System with MCA (simulations).

Algorithm	Membrane system	MCA
Sum of squares	1.9 µs	2.9 µs
Product of squares	2.3 µs	2.4 µs
Square + random	1.92 µs	2.92 µs
Cubic random	1.93 µs	3.93 µs
Square + random	1.92 µs	2.92 µs
NAND continuous	2.83 µs	2.87 µs
XOR continuous	2.72 µs	2.56 µs
Cubic random AND XOR	3.96 µs	4.01 µs
Square + random AND XOR	3.82 µs	3.52 µs
Cubic random CONTINUOS XOR	4.77 µs	3.99 µs

The analysis is very direct. The simulations are running in the same platform and just focuses in performance time based. All problems are considered simple problems due to the limitations of processing a complex problem with a complex set of aggregation rules which will jeopardize the accuracy of the analysis. Nevertheless, it is indicative to see that there is a variation in the performance when the level of complexity slightly increases which suggest that aggregation can be a good approach when the level of complexity increases.

6 Conclusions

Membrane computing has been growing since George Paun defined it in 1998. Since then new variations have been suggested to try to fit this model to new realities. The main goal for this unconventional paradigm is to improve the performance of the traditional algorithms due to the inherent limitation of the model. Simulations are still a big part of membrane computing and they are useful to extract right conclusions about the new model. In particular, this model is a great candidate to be applied to complex models that require an aggregated solution that is part of other sub solution whole super solutions as long as the defined rules in the MCA are followed. The aggregation factor that is linked to the minimal membrane cells is the component that complement the use membrane computing as a whole and as unite aggregated model. As the creation of this factor generates difficulties because it depends on the nature of the problem, it does not damage the performance during the execution as the factor is calculated in compiling time. New techniques to atomize the generation of λ as this could create a complete dynamic model that fully adjust to the problem and create the right MCA. The necessity of opening the line of research is out of question. The field is growing and new experiments are required. MCA systems are provided as a natural solution to upgrade the nature of membrane computing by not only taking advantage of the properties of the membrane cells but by the way these cells are aggregated. The future work will be involving complex problems in complex aggregated structures, so the analysis can be more relevant. Nevertheless, the evidence points out that aggregation is a natural solution to deal with complex problems that nowadays are being processed by conventional approaches such as backtracking or dynamic programming.

References

1. Adelman, L.M.: Molecular computation of solutions to combinatorial problems. Science **226**, 1021–1024 (1994)
2. Amir-Kroll, H., Sadot, A., Cohen, I.R., Harel, D.: GemCell: a generic platform for modelling multi-cellular. Theor. Comput. Sci. **391**, 276–290 (2008)
3. Arteta, A., Mingo, L.F., Castellanos, J.: An isomorphism based algorithm to solve complex problems. WSEAS Trans. Inf. Sci. Appl. **15**, 27–36 (2018)
4. Arteta, A., Mingo, L.F., Gomez, N.: New approach to optimize membrane systems. J. Bioinform. **1**(1), 1–6 (2014)
5. Arteta, A., Mingo, L.F., Gomez, N.: Membrane systems working with the P-factor: best strategy to solve complex problems. Adv. Sci. Lett. **19**(5), 1490–1495 (2012)

6. Arteta, A.: MEIA systems: membrane encrypted information application systems. Nat. Inf. Technol. Madr. Int. J. Inf. Theor. Appl. **19**(2), 103–109 (2012)
7. Arteta, A., Gomez, N., Gonzalo, R.: Solving diophantine equations with a parallel membrane computing model. Int. J. Inf. Models Anal. **1**, 220–225 (2012)
8. Arteta, A., Mingo, L.F., Gomez, N.: Solving complex problems with a bio-inspired model. Eng. Appl. Artif. Intell. **24**(6), 919–927 (2011)
9. Arteta, A., Fernández, L., Arroyo, F.: P-systems: study of randomness when applying evolution rules. In: International Book Series "Information Science and Computing, pp. 15–24 (2009)
10. Arteta, A., Goñi, A., Castellanos, J.: Analysis of P-systems under multiagents perspective. In: International Book Series "Information Science and Computing", pp. 117–128 (2009)
11. Angeleska, A., et al.: RNA-guided DNA assembly. J. Theor. Biol. **248**, 706–720 (2007)
12. Ardelean, I., et al.: A computational model for cell differentiation. BioSystems **76**(1–3), 169–176 (2004)
13. Balazsi, G., Barabasi, A.-L., Oltvai, Z.N.: Topological units of environmental signal processing in the transcriptional regulatory network of Escherichia coli. PNAS **102**(22), 7841–7846 (2005)
14. Bashor, C.J., Horwitz, A.A., Peisajovich, S.J., Lim, W.A.: Rewiring cells: synthetic biology as a tool to interrogate the organizational principles of living systems. Annu. Rev. Biophys. **39**, 515–537 (2010)
15. Blow, N.: Systems biology: untangling the protein web. Nature **460**, 415–418 (2009)
16. Bray, D.: Wetware: A Computer in Every Living Cell: The Computer in Every Living Cell. Yale University Press, New Haven (2009)
17. Brooks, R.A.: The relationship between matter and life. Nature **409**, 409–411 (2001)
18. Cao, H., Romero-Campero, F.J., Heeb, S., Camara, M., Krasnogor, N.: Evolving cell models for systems and synthetic biology. Syst. Synth. Biol. **4**(1), 55–84 (2010)
19. Carroll, S.B.: Endless Forms Most Beautiful. W.W. Norton, Chicago (2005)
20. Danchin, A.: Bacteria as computers making computers. FEMS Microbiol. Rev. **33**(1), 3–26 (2009)
21. Dasgupta, D. (ed.): Artificial Immune Systems and Their Applications. Springer, Heidelberg (1998). https://doi.org/10.1007/978-3-642-59901-9
22. Dassow, J., Mitrana, V.: On some operations suggested by genome evolution. In: Proceedings of the Second Pacific Symposium on Biocomputing, pp. 97–108 (1997)
23. Dassow, J., Mitrana, V., Salomaa, A.: Context-free evolutionary grammars and the structural language of nucleic acids. BioSystems **43**, 169–177 (1997)
24. Dassow, J., Mitrana, V., Salomaa, A.: Operations and language generating devices suggested by the genome evolution. Theor. Comput. Sci. **270**(1–2), 701–731 (2002)
25. Edelman, G.M.: Topobiology: An introduction to molecular embriology. Basic Books, New York (1988)
26. Ehrenfeucht, A., Harju, T., Petre, I., Prescott, D.M., Rozenberg, G.: Computation in Living Cells: Gene Assembly in Ciliate. Natural Computing Series. Springer, Heidelberg (2003). https://doi.org/10.1007/978-3-662-06371-2
27. Engelbrecht, A.: Fundamentals of Computational Swarm Intelligence. Wiley, Chichester (2005)
28. Federici, D., Downing, K.: Evolution and development of a multicellular organism: scalability, resilience, and neutral complexification. Artif. Life **12**(3), 381–409 (2006)
29. Freund, R., Martin-Vide, C., Mitrana, V.: On some operations suggested by gene assembly in ciliates. New Gener. Comput. **20**, 279–293 (2002)
30. de Frutos, J.A., Fernández, L., Luengo, C., Arteta, A.: Improving active rules performance in new P system communication architectures. Inf. Technol. Knowl. **4**(1), 3–18 (2010)

31. Arteta, A., Castellanos, A., Martinez, A.: Membrane computing: non deterministic technique to calculate extinguished multisets of objects. Int. J. Inf. Technol. Knowl. 4(1), 30–41 (2010)
32. de Frutos, J.A., Arroyo, F., Arteta, A.: Usefulness states in new P system communication architectures. In: Corne, D.W., Frisco, P., Păun, G., Rozenberg, G., Salomaa, A. (eds.) WMC 2008. LNCS, vol. 5391, pp. 169–186. Springer, Heidelberg (2009). https://doi.org/10.1007/978-3-540-95885-7_13
33. Han, J., et al.: Evidence for dynamically organized modularity in the yeast protein–protein interaction network. Nature 430, 88–93 (2004)
34. Haynes, K.A., Silver, P.A.: Eukaryotic systems broaden the scope of synthetic biology. JCB 187(5), 589–596 (2009)
35. Ho, L., Crabtree, G.R.: Chromatin remodelling during development. Nature 463, 474–484 (2010)
36. Holcombe, M., Bell, A.: Computational models of immunological pathways. In: Holcombe, M., Paton, R. (eds.) Information Processing in Cells and Tissues. Plenum Press, New York (1998)

Author Index

Printed in the United States
By Bookmasters